Springer Series in Solid-State Sciences

Volume 213

The Springer Series in Solid-State Sciences features fundamental scientific books prepared by leading and up-and-coming researchers in the field. They strive to communicate, in a systematic and comprehensive way, the basic principles as well as new developments in theoretical and experimental solid-state science.

We welcome submissions for monographs or edited volumes from scholars across this broad domain. Topics of current interest include, but are not limited to:

- Semiconductors and superconductors
- Quantum phenomena
- Spin physics
- Topological insulators
- Multiferroics
- Nano-optics and nanophotonics
- Correlated electron systems and strongly correlated materials
- Vibrational and electronic properties of solids
- Spectroscopy and magnetic resonance

Kashif Ammar Yasir · Wu-Ming Liu

Rydberg Atoms in Cavity

Springer

Kashif Ammar Yasir
Department of Physics
Zhejiang Normal University
Jinhua, Zhejiang, China

Wu-Ming Liu
Institute of Physics
Chinese Academy of Sciences
Beijing, China

Tsientang Institute for Advanced Study
Hangzhou, Zhejiang, China

ISSN 0171-1873 ISSN 2197-4179 (electronic)
Springer Series in Solid-State Sciences
ISBN 978-981-95-4570-4 ISBN 978-981-95-4571-1 (eBook)
https://doi.org/10.1007/978-981-95-4571-1

This Springer imprint is published by the registered company Springer Nature Singapore Pte Ltd.
The registered company address is: 152 Beach Road, #21-01/04 Gateway East, Singapore 189721, Singapore

This book is lovingly dedicated to my parents, Saira Afzal and Muhammad Afzal, whose boundless love, wisdom, and unwavering guidance have illuminated every path of my life. Their faith in me has been the strength behind all my endeavors, and their sacrifices the silent pillars on which my journey rests.

To my wife, Madeeha, I owe profound gratitude for her steadfast support and companionship. In moments of triumph and in times of trial, she has been my anchor offering patience, encouragement, and serenity that have kept me balanced and inspired.

Preface

Rydberg atoms in cavities represent a uniquely powerful and versatile platform for exploring strongly interacting quantum systems, enabling deterministic quantum gates, scalable entanglement generation, and photonic nonlinearities at the single-photon level. Their exaggerated atomic properties, such as large electric dipole moments, long radiative lifetimes, and extreme sensitivity to external fields combined with the engineered light-matter interaction afforded by optical and microwave resonators, place them at the forefront of modern quantum optics, quantum simulation, and quantum information science.

This book offers a comprehensive and self-contained account of the physics of Rydberg atoms and their integration within the framework of cavity quantum electrodynamics (QED) and hybrid quantum systems. Our goal is to equip the reader with both the foundational understanding and the advanced technical tools required to navigate and contribute to the current landscape of Rydberg-based quantum science. Beginning with the preparation and characterization of Rydberg states, we explore their core interaction mechanisms including dipole blockade, Förster Resonance, and tunable long-range van der Waals and dipole-dipole interactions which underlie many of the novel phenomena observed in this field.

We then examine the interplay between strong atomic interactions and spatially confined electromagnetic fields, highlighting how cavity QED architectures enhance, modify, and in some cases entirely redefine light-matter coupling. Particular emphasis is placed on the role of high-finesse optical and superconducting microwave cavities in mediating photon-photon interactions, generating collective excitations, and facilitating the emergence of strongly correlated quantum states. These fundamental concepts are systematically developed and then extended to advanced topics such as cavity-assisted entanglement generation, deterministic quantum gates, quantum non-demolition measurements, and nonclassical light sources.

The book also addresses recent breakthroughs and emerging directions in the field, including single-photon switches and transistors, Rydberg electromagnetically induced transparency (EIT), Rydberg polaritons, and programmable quantum registers realized with tweezer-trapped atom arrays coupled to cavity fields. Theoretical models are presented with rigor and clarity, complemented by discussions of experimental implementations drawn from state-of-the-art research. Throughout the text, we place strong emphasis on physical intuition, analytical techniques, and the interplay between theoretical constructs and experimental feasibility. This volume is intended for graduate students, postdoctoral researchers, and advanced undergraduates in atomic physics, quantum optics, quantum information, and related disciplines. A working knowledge of quantum mechanics and classical electrodynamics is assumed, but no prior familiarity with Rydberg atoms or cavity QED is required.

We hope this book serves not only as an accessible entry point for newcomers but also as a lasting reference for those actively engaged in cutting-edge research in quantum science and technology. The rich convergence of strong interactions, coherent control, and cavity-enhanced dynamics in Rydberg systems continues to uncover new physics and inspire novel applications an exciting frontier we invite the reader to explore with us.

Jinhua, China — Kashif Ammar Yasir
Hangzhou/Beijing, China — Wu-Ming Liu

Acknowledgements First and foremost, I bow in gratitude to Almighty Allah for granting me the strength, courage, and resilience to face the complexities of life and for guiding me through every step of this journey.

I am deeply indebted to my parents, Saira Afzal and Muhammad Afzal, whose unconditional love, care, and unwavering support have been the foundation of my life. To my wife, Madeeha, I extend heartfelt thanks for being a constant source of stability and continuity, whose patience and strength have carried me through challenging times. I also wish to acknowledge my three little angels, Rayyan, Imaan, and Rumman, who bring endless joy, hope, and inspiration into my life.

My sisters—Saima, Asma, Nida, and Fareeha—have always surrounded me with love, encouragement, and warmth, for which I am forever grateful. To my brother, Fahad, I owe a very special mention. He was not only my companion but truly my right arm, a source of strength and support. His absence has left me incomplete, yet his memory continues to inspire and guide me.

My utmost gratitude goes to Prof. Wu-Ming Liu, whose sincerity, wisdom, and constant encouragement I will always cherish and remember. His guidance has been invaluable to both my academic growth and personal development.

I am also thankful to my friends and colleagues, whose companionship and encouragement have made this journey more meaningful.

I acknowledge the support of Research Fund for International Young Scientists by NSFC under grant Nos. KYZ04Y22050, KYZ04Y22165, Zhejiang Normal University research funding under grant No. ZC304021914 and Zhejiang province postdoctoral research project under grant number ZC304021952. This work was also supported by National Key R&D Program of China under grants No. 2021YFA1400900, 2021YFA0718300, 2021YFA1400243, NSFC under grants Nos. 61835013, Space Application System of China Manned Space Program.

Competing Interests The authors have no competing interests to declare that are relevant to the content of this manuscript.

Contents

Chapter 1
Introduction

Abstract This chapter introduces the motivation for studying Rydberg atoms in the context of cavity quantum electrodynamics (QED). It outlines the unique properties of highly excited Rydberg states, the role of strong and tunable interactions, and the importance of coupling Rydberg ensembles to confined electromagnetic modes. The discussion sets the stage for exploring fundamental phenomena such as blockade physics, quantum nonlinear optics, and cavity-mediated phase transitions, as well as their applications to emerging quantum technologies.

In 1913, the Danish physicist Niels Bohr proposed a pioneering model of the atom now famously known as Bohr's atomic model which provided one of the earliest quantum explanations for atomic structure and spectral lines [1]. This model introduced the concept of quantized electron orbits around a nucleus, where electrons could only occupy specific energy levels and would emit or absorb energy in discrete amounts when transitioning between these levels. Although Bohr's model was eventually superseded by the more complete and sophisticated formulations of quantum mechanics, it laid the groundwork for much of modern atomic physics. Interestingly, certain ideas from Bohr's model continued to inspire theoretical and experimental investigations, especially in regimes that bridge classical and quantum domains. One such regime is the study of atoms in highly excited states an area that has matured into a significant subfield known as Rydberg atomic physics.

In atomic physics, highly excited atoms those in which an electron is promoted to a very high principal quantum number display properties that differ drastically from ground-state or low-excitation atoms. These atoms occupy a middle ground between classical mechanics and quantum behavior and offer a rich playground for exploring the dynamics of electron motion, long-range interactions, and light-matter coupling. Atoms with such large excitation levels are termed Rydberg atoms, named after the Swedish spectroscopist Johannes Rydberg, who in the late 19th century developed an empirical formula to describe the spectral lines of hydrogen-like systems with an empirical parametric fit, which interesting later was defined by Bohr model as Rydberg constant in mathematical formalism [2–5]. These atoms are remarkable not

K. A. Yasir and W.-M. Liu, *Rydberg Atoms in Cavity*, Springer Series in Solid-State Sciences 213, https://doi.org/10.1007/978-981-95-4571-1_1

only because of their extreme size which scales as n^2 (hydrogenic Rydberg atoms, while atoms like alkalis can deviate slightly due to quantum defect effects) but also due to the extremely weak binding of the outer electron to the atomic core. This outermost electron, often referred to as the Rydberg electron, is highly sensitive to external electromagnetic fields and perturbations, making Rydberg atoms uniquely suited for precision measurements and studies of fundamental interactions [6–11].

The creation of Rydberg atoms can be achieved through several experimental techniques. One widely used approach (especially for astrophysical or plasma environments) involves the recombination of free electrons with positively charged ions, accompanied by photon emission that places the atom in a highly excited state. Another common method uses high-intensity coherent laser fields to excite ground-state atoms to specific Rydberg levels. Additionally, collisional excitation during controlled interatomic collisions can also lead to Rydberg formation [12–20]. It is important to emphasize that not all excited atoms are considered Rydberg atoms; generally, the principal quantum number must exceed $n > 10$ for the atom to exhibit characteristic Rydberg behavior, although this is more of a conventional threshold than a strict physical boundary, where Rydberg-like properties (large radius, high polarizability) become dominant. Remarkably, in the vastness of interstellar space, natural Rydberg atoms with quantum numbers exceeding $n \approx 1000$ have been observed, particularly in carbon atoms [21–24]. These atoms, due to their extreme sensitivity to weak electromagnetic fields, play a crucial role in astrophysical spectroscopy and magnetic field measurements across astronomical distances.

Beyond interstellar space, Rydberg atoms can also be generated under laboratory conditions involving strong magnetic or electric fields. For example, experiments involving antimatter plasmas, such as those conducted at CERN, have demonstrated the formation of Rydberg positronium atoms through electron-positron recombination in magnetic traps [25, 26]. Similarly, in the realm of cold atom physics, ultra-cold atoms trapped in magneto-optical traps or optical lattices can be coherently excited to Rydberg states using narrowband lasers [27–29]. These experimental platforms provide a high degree of control over atomic motion and interactions, opening new opportunities for investigating quantum phenomena such as entanglement, coherence, and many-body effects in Rydberg systems.

Parallel to the development of Rydberg physics, the quantum theory of light underwent its own profound transformation, reshaping our understanding of light-matter interactions at a fundamental level. The notion that electromagnetic radiation is quantized that it is composed of discrete energy packets called photons was first introduced by Max Planck in 1899 in his successful attempt to resolve the ultraviolet catastrophe associated with black-body radiation. Planck's quantization of energy, though initially a mathematical trick, laid the conceptual groundwork for quantum theory. This idea gained physical significance when Albert Einstein, in 1905, proposed the photon model of light to explain the photoelectric effect, wherein electrons are ejected from a metallic surface upon exposure to ultraviolet light. This work provided compelling evidence against the purely wave-based theory of light and introduced the concept of quantized electromagnetic energy as a physical reality. The quantization of radiation became a cornerstone of modern physics, fundamentally influencing Niels Bohr's model of the hydrogen atom and the quantized nature of atomic spectra. These historical milestones collectively established the discrete nature of both energy and radiation, setting the stage for the development of quantum electrodynamics (QED), which unifies the quantization of light and matter into a coherent framework [30–37]. This foundational understanding became especially crucial in exploring complex systems like Rydberg atoms, where quantization plays a dominant role in long-range interaction and state evolution.

A key insight of Einstein's theoretical framework was the phenomenon of spontaneous emission a process in which an excited atom emits a photon even in the absence of any stimulating external electromagnetic field. This inherently quantum process is rooted in the coupling of atomic states to vacuum field fluctuations and is responsible for the inevitable decay of excited states to lower energy configurations. Spontaneous emission not only ensures thermal equilibrium in radiative systems but also reveals the stochastic nature of quantum evolution [38–41]. When analyzed from the perspective of quantum optics a field that rigorously describes light as a quantized field with well-defined statistical and coherence properties these interactions transcend the simplistic energy transfer viewpoint. They encompass quantum noise, entanglement generation, and the emergence of coherence phenomena. For instance, the photon number distribution, second-order coherence functions, and quantum state collapse are all critical elements that govern light-matter interaction in both free space and confined systems. When Rydberg atoms, with their exaggerated dipole moments and long coherence times, interact with such quantized fields, they exhibit rich dynamics that are both tunable and highly nonclassical [42–45]. These attributes make QED, particularly in the cavity setting, a fertile ground for generating and studying entangled states, nonlinear optical phenomena, and quantum logical operations [46–49].

A particularly fascinating and subtle feature of quantum electrodynamics is the concept of vacuum fluctuations microscopic field disturbances that persist even in the absence of classical electromagnetic radiation. These fluctuations are not merely theoretical curiosities; they have measurable consequences, such as the Lamb shift and the Casimir effect. When an atom is placed inside a high-finesse optical cavity, the boundary conditions of the electromagnetic field are altered, leading to a modified vacuum mode structure and altered spontaneous emission dynamics. In essence, the cavity acts as a mode filter, enhancing or suppressing certain field modes and thereby tailoring the electromagnetic environment of the atom. This phenomenon is known as the Purcell effect and plays a pivotal role in cavity QED experiments. A prototypical experimental configuration includes two highly reflective mirrors forming an optical resonator, where a beam of atoms often cooled to ultracold temperatures and prepared in well-defined quantum states is injected into the cavity mode. These atoms, frequently chosen to be Rydberg states due to their long radiative lifetimes and large interaction cross-sections, interact coherently with the confined photons in the cavity. The resulting dynamics are highly sensitive to the detuning, mode volume, and coupling strength, allowing precise control over quantum interference and coherent population trapping [35, 50–52]. Figure 1.1 schematically depicts such a

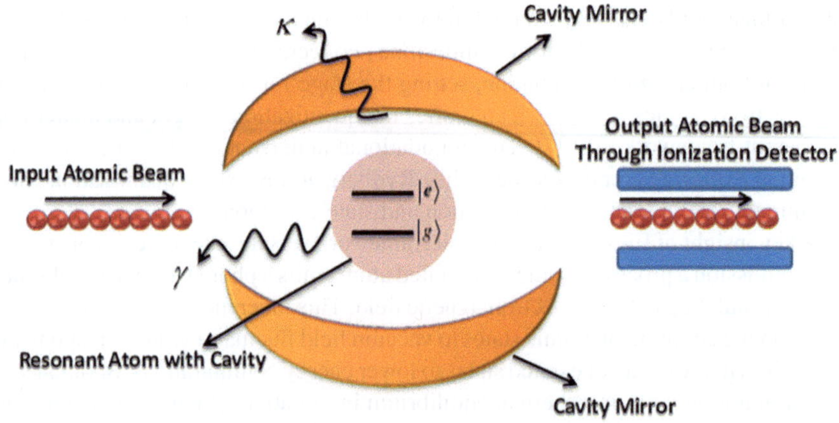

Fig. 1.1 Schematic diagram of an optical cavity with two-level atoms. A beam of two-level atoms interacts with intra-cavity optical mode which is resonant with atomic transition frequency and ejects a photon into the cavity. Here κ is the photon decay rate of intra-cavity field energy and parameter γ accounts for non-resonant decay of atomic mode caused by the spontaneous emission. The shaded region inside the cavity represents resonant interaction of atoms with cavity mode. The ionization detector measures the ionization of atomic beam when it takes an exits from the cavity

setup, where the interplay between Rydberg-state-induced nonlinearities and cavity confinement gives rise to tunable light-matter hybrid states, including Rydberg polaritons and photon blockade phenomena.

Despite the conceptual clarity of cavity QED, experimental realization of strong coupling between atoms and quantized cavity fields faced formidable technical challenges, particularly during its formative years in the 1980s. The central criterion for achieving the strong coupling regime where the coherent interaction strength g between the atom and the cavity field exceeds both the atomic decay rate γ and the cavity loss rate κ required innovations in cavity fabrication, atom preparation, and noise isolation. Early attempts to reach this regime were hindered by photon losses and decoherence. As Serge Haroche vividly described it, achieving this regime was akin to the "Holy Grail" of quantum optics [53]. A major difficulty lay in maintaining long photon lifetimes, necessitating ultra-high reflective coatings and precise alignment of the cavity mirrors to achieve high quality factors $Q = \omega/\kappa$, where ω is the cavity resonance frequency. The success of research groups such as those at the Max Planck Institute marked a pivotal moment in quantum optics. Utilizing advances in atomic beam techniques, cryogenic cooling, and piezoelectric stabilization, these researchers demonstrated unambiguous signatures of vacuum Rabi splitting and coherent oscillations between cavity photons and atomic dipoles [54–65]. These observations confirmed the theoretical predictions of cavity QED and opened new avenues in quantum information science. Subsequent experiments incorporated Rydberg atoms to explore photon-mediated interaction networks, deterministic entanglement protocols, and exotic phases of light, firmly establishing cavity QED with Rydberg atoms as a cornerstone platform for modern quantum technologies [66–70].

Rydberg atoms inside optical cavities form an especially powerful hybrid platform for exploring these phenomena. The large dipole moments and strong polarizabilities of Rydberg states make them highly sensitive to both cavity photons and external perturbations [71–78]. When ensembles of Rydberg atoms interact with a driven cavity mode, long-range dipole-dipole interactions become significant, giving rise to collective behavior and photon-mediated correlations [79–84]. In contrast to typical dipolar systems, the strength, range, and angular dependence of these interactions can be dynamically tuned via the cavity field parameters, thus offering a high degree of experimental control.

A particularly noteworthy phenomenon in this context is the Rydberg blockade, where excitation of one atom to a Rydberg state shifts the energy levels of neighboring atoms out of resonance, thereby suppressing further excitations within a given blockade radius [85–90]. This collective excitation behavior allows the ensemble to behave effectively as a two-level system, which is highly beneficial for quantum information protocols. In other words, within each blockade radius, only a single excitation is allowed, mimicking a collective two-level system. The Rydberg blockade has become a cornerstone in the design of quantum gates, single-photon sources, and entangled state preparation.

Beyond quantum optics, Rydberg atoms have demonstrated utility across multiple domains of physics. In astrophysics, they assist in mapping interstellar media and tracking cosmic magnetic fields [91–94]. In high-energy physics, their interactions provide insight into particle-antiparticle dynamics and strong-field effects [95–98]. Plasma physicists study them to understand radiative recombination, collisional excitation, and energy transport processes [99–104]. Rydberg atoms also offer experimental platforms for probing quantum chaos, classical-quantum correspondence, and the onset of decoherence [105–113]. Crucially, in the field of quantum information science, Rydberg atoms particularly when coupled with cavity modes serve as versatile building blocks for implementing scalable quantum networks, quantum memories, and logic operations [114–126].

In this book, we aim to present a detailed and comprehensive account of the interaction between Rydberg atoms and optical cavities, progressing from fundamental theoretical principles to modern experimental realizations and emerging applications in quantum technology. In Chap. 2, we provide a detailed exposition of the properties of Rydberg atoms and molecules, as well as their interactions with electric and magnetic fields. We discuss the Rydberg blockade and the mechanisms of long-range dipole-dipole interactions. Chapter 3 is dedicated to the fundamental processes of absorption and emission inside the cavity and the formation of photon number states. We also review the realization of atomic masers through cavity-assisted Rydberg dynamics. In Chap. 4, we explore nonlinear optical effects, including electromagnetically induced transparency and Rydberg polaritons. The Chap. 5 covers the quantum phase transitions driven by cavity in Rydberg atoms. Chapter 6 focuses on dynamic properties such as optical bistability, stability analysis, and hybrid optomechanics involving Rydberg systems. Chapter 7 presents recent experimental and theoretical developments in applying Rydberg atoms to quantum information tasks, including entanglement generation and quantum gate implementation. The book concludes with a discussion on future directions in Chap. 8, highlighting the ongoing potential of this rapidly evolving field.

References

1. N. Bohr, On the constitution of atoms and molecules. Philos. Mag. **26**, 1 (1913)
2. T.F. Gallagher, *Rydberg Atoms* (Cambridge University Press, 1994)
3. E.J. Connerade, *Highly Excited Atoms* (Cambridge University Press, 1998)
4. M. Born, *The Mechanics of the Atom* (Federick Ungar Publishing Co., New York, 1960)
5. R.F. Stebbings, F.B. Dunning (eds.), *Rydberg States of Atoms and Molecules* (Cambridge University Press, 1983)
6. H.A. Bethe, E.E. Salpeter, *Quantum Mechanics of One- and Two-Electron Atoms* (Springer-Verlag, Berlin, 1957)
7. L.D. Landau, E.M. Lifshitz, *Quantum Mechanics* (Pergamon Press, London, 1958)
8. L. Pauling, E.B. Wilson, *Introduction to Quantum Mechanics* (McGraw-Hill Inc., 1935)
9. E. Merzbacher, *Quantum Mechanics*, 2nd ed. (John Wiley and Sons, 1970)
10. J.L. Powell, B. Crasemann, *Quantum Mechanics* (Addison-Wesley Pub. (Sd), 1961)
11. D. Kleppner, M.G. Littman, M.L. Zimmerman, Rydberg atoms in strong field, in *Rydberg States of Atoms and Molecules*, eds. by R.F. Stebbings, F.B. Dunning (Cambridge University Press, 1983), pp. 73–116
12. C. Fabre, S. Haroche, Spectroscopy of one- and two-electron Rydberg atoms, in *Rydberg States of Atoms and Molecules*, eds. by R.F. Stebbings, F.B. Dunning (Cambridge University Press, 1983), pp. 117–164
13. M. Matsuzawa, in *Rydberg States of Atoms and Molecules*, eds. by R.F. Stebbings, F.B. Dunning (Cambridge University Press, 1983), pp. 267–314
14. D. Dai et al., State-to-state dynamics of high-n Rydberg H-atom scattering with D_2. Phys. Rev. Lett. **95**, 013201 (2005)
15. K.B. MacAdam, D.A. Crosby, R. Rolfes, n Dependence of l-changing collisions between He^+ ions and Na. Phys. Rev. Lett. **44**, 980 (1980)
16. W.D. Phillips, Nobel lecture: laser cooling and trapping of neutral atoms. Rev. Mod. Phys. **70**, 721 (1998)
17. S. Chu, Nobel lecture: the manipulation of neutral particles. Rev. Mod. Phys. **70**, 685 (1998)
18. V. Bendkowsky et al., Observation of ultralong-range Rydberg molecules. Nature **458**, 1005–1008 (2009)
19. C. Vadla, V. Horvatic, K. Niemax, Line-broadening studies of excited diatomic homoatomic and heteroatomic Rydberg molecules formed by potassium, rubidium, and cesium atoms. Phys. Rev. A **80**, 052506 (2009)
20. I. Lesanovsky, M. Müller, P. Zoller, Trap-assisted creation of giant molecules and Rydberg-mediated coherent charge transfer in a Penning trap. Phys. Rev. A **79**, 010701(R) (2009)
21. S.V. Stepkin, A.A. Konovalenko, N.G. Kantharia, N.U. Shankar, Radio recombination lines from the largest bound atoms in space. Mon. Not. R. Astron. Soc. **374**, 852 (2007)
22. A.P. Sarma, E. Momjian, T.H. Troland, R.M. Crutcher, Very large array HI Zeeman observations of NGC 1275 (Perseus a). Astron. J. **130**, 2566 (2005)
23. A. Dalgarno, Rydberg atoms in astrophysics, in *Rydberg States of Atoms and Molecules*, eds. by R.F. Stebbings, F.B. Dunning (Cambridge University Press, 1983), pp. 1–30
24. T.H. Troland, Astronomical polarimetry, current status and future directions. ASP Conf. Ser. **343**, 64 (2005)
25. M. Amoretti et al., Production and detection of cold Antihydrogen atoms. Nature **419**, 456 (2002)

26. G. Gabrielse et al., Background-free observation of cold Antihydrogen with field-ionization analysis of its states. Phys. Rev. Lett. **89**, 213401 (2002)
27. G. Gwinner et al., Influence of magnetic fields on electron-ion recombination at very low energies. Phys. Rev. Lett. **84**, 4822 (2000)
28. J.H. Choi, J.R. Guest, A.P. Povilus, E. Hansis, G. Raithel, Magnetic trapping of long-lived cold Rydberg atoms. Phys. Rev. Lett. **95**, 243001 (2005)
29. R.H. Garstang, Atoms in high magnetic-fields. Rep. Prog. Phys. **40**, 105 (1977)
30. A. Einstein, B. Podolsky, N. Rosen, Can quantum-mechanical description of physical reality be considered complete? Phys. Rev. **47**, 777 (1935)
31. E. Schrödinger, Naturwissenschaften **23**, 807 (1935)
32. R.P. Feynman, *QED: The Strange Theory of Light and Matter* (Princeton University Press, 1985)
33. L. De Broglie, *Recherches sur la theorie des quanta (Research on quantum theory)* (Wiley-Interscience, France, 1925)
34. N. Bohr, Discussion with Einstein on epistemological problems in atomic physics, in *Albert Einstein, Philosopher-Scientist*, ed. by P.A. Shilpp (Harper, New York, 1949)
35. S. Haroche, Nobel lecture: controlling photons in a box and exploring the quantum to classical boundary. Rev. Mod. Phys. **85**, 1083 (2013)
36. G. Greenstein, A.G. Zajonc, *The Quantum Challenge: Modern Research on the Foundations of Quantum Mechanics* (Jones and Bartlett Publishers, London, 1997)
37. T.D. Newton, E.P. Wigner, Localized states for elementary systems. Rev. Mod. Phys. **21**, 400 (1949)
38. A. Ashkin, Acceleration and trapping of particles by radiation pressure. Phys. Rev. Lett. **24**, 156 (1970)
39. A. Ashkin, Atomic-beam detection by resonance-radiation pressure. Phys. Rev. Lett. **25**, 1321 (1970)
40. Y.H. Shih, C.O. Alley, New type of Einstein–Podolsky–Rosen–Bohm experiment using pairs of light quanta produced by optical parametric down conversion. Phys. Rev. Lett. **61**, 2921 (1988)
41. M. Genovese, Research on hidden variable theories: a review of recent progresses. Phys. Rep. **413**, 319 (2005)
42. K. Hammerer, A.S. Sørensen, E.S. Polzik, Quantum interface between light and atomic ensembles. Rev. Mod. Phys. **82**, 1041 (2010)
43. M.O. Scully, M.S. Zubairy, *Quantum Optics* (Cambridge University Press, 1997)
44. W.P. Schleich, *Quantum Optics in Phase Space* (Wiley-VCH Verlag Berlin GmbH, Berlin, 2001)
45. V. Vedral, *Modern Foundations of Quantum Optics* (Imperial College Press, London, 2005)
46. S. Haroche, J.-M. Raimond, *Exploring the Quantum: Atoms* (Oxford University Press, Cavities and Photons, 2006)
47. D.F. Walls, G.J. Milburn, *Quantum Optics* (Springer, 1994)
48. R.P. Feynman, *Quantum Electrodynamics* (New ed.) (Westview Press, 1998)
49. J. Schwinger, *Selected Papers on Quantum Electrodynamics* (Dover Publications, 1958)
50. A.P. Kazantsev, G.I. Surdutovich, V.P. Yakovle, *Mechanical Action of Light on Atoms* (World Scientific Pub. Co., Singapore, 1990)
51. R.J. Cook, A.F. Bernhardt, Detection of atoms by a resonant standing electromagnetic wave. Phys. Rev. A **18**, 2533 (1978)
52. P.J. Martin, P.L. Gould, B.G. Oldaker, A.H. Miklich, D.E. Pritchard, Diffraction of atoms moving through a standing light wave. Phys. Rev. A **36**, 2495 (1987)
53. S. Haroche, J.M. Raimond, Cavity quantum electrodynamics. Sci. Am. **268**, 54–60 (1993)
54. C.J. Hood, M.S. Chapman, T.W. Lynn, H.J. Kimble, Real-time cavity QED with single atoms. Phys. Rev. Lett. **80**, 4157 (1998)
55. T. Pellizzari, S.A. Gardiner, J.I. Cirac, P. Zoller, Decoherence, continuous observation, and quantum computing: A cavity QED model. Phys. Rev. Lett. **75**, 3788 (1995)

56. J. Ye, D.W. Vernooy, H.J. Kimble, Trapping of single atoms in cavity QED. Phys. Rev. Lett. **83**, 4987 (1999)
57. J. McKeever et al., State-insensitive cooling and trapping of single atoms in an optical cavity. Phys. Rev. Lett. **90**, 133602 (2003)
58. M.S. Chapman et al., Photon scattering from atoms in an atom interferometer: coherence lost and regained. Phys. Rev. Lett. **75**, 3783 (1995)
59. S. Dürr, T. Nonn, G. Rempe, Origin of quantum-mechanical complementarity probed by a which-way experiment in an atom interferometer. Nature **395**, 33 (1998)
60. C.J. Hood, T.W. Lynn, A.C. Doherty, A.S. Parkins, H.J. Kimble, The atom-cavity microscope: single atoms bound in orbit by single photons. Science **287**, 1447 (2000)
61. S. Brattke, B.T.H. Varcoe, H. Walther, Generation of photon number states on demand via cavity quantum electrodynamics. Phys. Rev. Lett. **86**, 3534 (2001)
62. K.J. Vahala, Optical microcavities. Nature **424**, 839 (2003)
63. K.M. Birnbaum et al., Photon blockade in an optical cavity with one trapped atom. Nature **436**, 87–90 (2005)
64. S. Gleyzes et al., Quantum jumps of light recording the birth and death of a photon in a cavity. Nature **446**, 297–300 (2007)
65. C. Guerlin et al., Progressive field-state collapse and quantum non-demolition photon counting. Nature **448**, 889–893 (2007)
66. S.M. Dutra, *Cavity Quantum Electrodynamics: The Strange Theory of Light in a Box* (John Wiley & Sons Inc, New Jersey, 2005)
67. H. Walther, B.T.H. Varcoe, B.-G. Englert, T. Becker, Cavity quantum electrodynamics. Rep. Prog. Phys. **69**, 1325 (2006)
68. S. Haroche, Entanglement, decoherence and the quantum-classical boundary. Phys. Today **51**, 36 (1998)
69. Y. Aharonov, M.S. Zubairy, Time and the quantum: erasing the past and impacting the future. Science **307**, 875 (2005)
70. G. Nogues, A. Rauschenbeutel, S. Osnaghi, M. Brune, J.M. Raimond, S. Haroche, Seeing a single photon without destroying it. Nature **400**, 239–242 (1999)
71. P. Filipovicz, P. Meystre, G. Rempe, H. Walther, Rydberg atoms. Opt. Acta Int. J. Opt. **32**, 1105–1123 (1985)
72. A. Joshi, R.R. Puri, Quantum electrodynamics of a Rydberg atom making two-photon transitions in the binomial state of the field in a lossless cavity. J. Mod. Opt. **36**, 215–224 (1989)
73. C. Cohen-Tannoudji, Nobel lecture: manipulating atoms with photons. Rev. Mod. Phys. **70**, 707 (1998)
74. W.D. Phillips, Nobel lecture: laser cooling and trapping of neutral atoms. Rev. Mod. Phys. **70**, 721 (1998)
75. S. Deléglise et al., Reconstruction of non-classical cavity field states with snapshots of their decoherence. Nature **455**, 510–514 (2008)
76. J.M. Fink et al., Climbing the Jaynes-Cummings ladder and observing its nonlinearity in a cavity QED system. Nature **454**, 315–318 (2008)
77. L.S. Bishop et al., Nonlinear response of the vacuum Rabi resonance. Nat. Phys. **5**, 105–109 (2009)
78. H.J. Kimble, The quantum internet. Nature **453**, 1023–1030 (2008)
79. P. Hyafil et al., Coherence-preserving trap architecture for long-term control of giant Rydberg atoms. Phys. Rev. Lett. **93**, 103001 (2004)
80. K. Singer et al., Suppression of excitation and spectral broadening induced by interactions in a cold gas of Rydberg atoms. Phys. Rev. Lett. **93**, 163001 (2004)
81. T.C. Liebisch, A. Reinhard, P.R. Berman, G. Raithel, Atom counting statistics in ensembles of interacting Rydberg atoms. Phys. Rev. Lett. **95**, 253002 (2005)
82. A. Reinhard et al., Rydberg–Rydberg collisions: resonant enhancement of state mixing and Penning ionization. Phys. Rev. Lett. **100**, 123007 (2008)
83. K. Singer, J. Stanojevic, M. Weidemüller, R. Côté, Long-range interactions between alkali Rydberg atom pairs correlated to the ns- ns, np-np and nd-nd asymptotes. J. Phys. B **38**, S295 (2005)

84. W.R. Anderson, J.R. Veale, T.F. Gallagher, Resonant dipole-dipole energy transfer in a nearly frozen Rydberg gas. Phys. Rev. Lett. **80**, 249 (1998)
85. C. Ates, T. Pohl, T. Pattard, J.M. Rost, Antiblockade in Rydberg excitation of an ultracold lattice gas. Phys. Rev. Lett. **98**, 023002 (2007)
86. R. Heidemann et al., Evidence for coherent collective Rydberg excitation in the strong blockade regime. Phys. Rev. Lett. **99**, 163601 (2007)
87. D. Tong et al., Local blockade of Rydberg excitation in an ultracold gas. Phys. Rev. Lett. **93**, 063001 (2004)
88. F. Robicheaux, J.V. Hernández, Many-body wave function in a dipole blockade configuration. Phys. Rev. A **72**, 063403 (2005)
89. I. Mourachko, D. Comparat, F. de Tomasi, A. Fioretti, P. Nosbaum, V.M. Akulin, P. Pillet, Many-body effects in a frozen Rydberg gas. Phys. Rev. Lett. **80**, 253 (1998)
90. T. Wilk, A. Gaëtan, C. Evellin, J. Wolters, Y. Miroshnychenko, P. Grangier, A. Browaeys, Entanglement of two individual neutral atoms using Rydberg blockade. Phys. Rev. Lett. **104**, 010502 (2010)
91. S.A. Edelstein, T.F. Gallagher, in *Advances in Atomic and Molecular Physics*. ed. by D. Bates, B. Bederson (Academic Press, New York, 1985)
92. Y.N. Gnedin et al., Rydberg atoms in astrophysics. New Astron. Rev. **53**, 259–265 (2009)
93. A.A. Mihajlov, M.S. Dimitrijevic, Z. Djuric, Collisional He-$He^*(n)$ chemi-ionization and dielectronic He-He^+-e and He_2^+-e recombination: differential and total reaction rate coefficients. Phys. Scr. **53**, 159 (1996)
94. M. Kilic et al., Mystery deepens: spitzer observations of cool white dwarfs. Astrophys. J. **642**, 2 (2006)
95. E. Widmann, Testing CPT with antiprotonic helium and antihydrogen—the ASACUSA experiment at CERN-AD. Nucl. Phys. A **752**, 87 (2005)
96. A. Kellerbauer et al., Proposed antimatter gravity measurement with an antihydrogen beam. Nucl. Instrum. Methods B **266**, 351 (2008)
97. R. Bluhm, V.A. Kostelecká, N. Russell, CPT and Lorentz tests in hydrogen and antihydrogen. Phys. Rev. Lett. **82**, 2254 (1999)
98. G. Gabrielse et al., First capture of antiprotons in a Dkiloelectronvolt source. Phys. Rev. Lett. **57**, 2504 (1986)
99. T.C. Killian, M.J. Lim, S. Kulin, R. Dumke, S.D. Bergeson, S.L. Rolston, Formation of Rydberg atoms in an expanding ultracold neutral plasma. Phys. Rev. Lett. **86**, 3759 (2001)
100. J.L. Roberts, C.D. Fertig, M.J. Lim, S.L. Rolston, Electron temperature of ultracold plasmas. Phys. Rev. Lett. **92**, 253003 (2004)
101. C.E. Simien et al., Using absorption imaging to study ion dynamics in an ultracold neutral plasma. Phys. Rev. Lett. **92**, 143001 (2004)
102. R.G. Greaves, M.D. Tinkle, C.M. Surko, Creation and uses of positron plasmas. Phys. Plasmas **1**, 1439 (1994)
103. L.V. Jørgensen et al., New source of dense, cryogenic positron plasmas. Phys. Rev. Lett. **95**, 025002 (2005)
104. G. Gabrielse et al., Antihydrogen production within a Penning-Ioffe trap. Phys. Rev. Lett. **100**, 113001 (2008)
105. D.J. Reilly, Engineering the quantum-classical interface of solid-state qubits. npj Quantum Inf. **1**, 15011 (2015)
106. K. Hammerer, A.S. Sørensen, E.S. Polzik, Quantum interface between light and atomic ensembles. Rev. Mod. Phys. **82**, 1041 (2010)
107. L. Krache, M. Maamache, Y. Saadi, Classical-quantum interface of a particle in a time-dependent linear potential. Int. J. Theor. Phys. **49**, 936 (2010)
108. S. Longhi, Classical and quantum interference in multiband optical Bloch oscillations. Phys. Rev. B **79**, 245109 (2009)
109. R.Z. Vered et al., Classical-to-quantum transition with broadband four-wave mixing. Phys. Rev. Lett. **114**, 063902 (2015)

110. R. Sawant et al., Nonclassical paths in quantum interference experiments. Phys. Rev. Lett. **113**, 120406 (2015)
111. M.C. Gutzwiller, *Chaos in Classical and Quantum Mechanics* (Springer-Verlag, New York, 1990)
112. A.V. Andreev, O. Agam, B.D. Simons, B.L. Altshuler, Quantum chaos, irreversible classical dynamics, and random matrix theory. Phys. Rev. Lett. **76**, 3947 (1996)
113. F. Saif, Classical and quantum chaos in atom optics. Phys. Rep. **419**, 207 (2005)
114. M. Saffman, T.G. Walker, K. Mølmer, Quantum information with Rydberg atoms. Rev. Mod. Phys. **82**, 2313 (2010)
115. M.A. Nielsen, I.L. Chuang, *Quantum Computation and Quantum Information* (Cambridge University Press, Cambridge, UK, 2000)
116. K. Bergmann, H. Theuer, B.W. Shore, Coherent population transfer among quantum states of atoms and molecules. Rev. Mod. Phys. **70**, 1003 (1998)
117. M.D. Lukin et al., Dipole blockade and quantum information processing in mesoscopic atomic ensembles. Phys. Rev. Lett. **87**, 037901 (2001)
118. E. Brion, K. Mølmer, M. Saffman, Quantum computing with collective ensembles of multi-level systems. Phys. Rev. Lett. **99**, 260501 (2007)
119. I. Bloch, Quantum coherence and entanglement with ultracold atoms in optical lattices. Nature **453**, 1016 (2008)
120. R. Blatt, D. Wineland, Entangled states of trapped atomic ions. Nature **453**, 1008 (2008)
121. M.T. Frey, S.B. Hill, K.A. Smith, F.B. Dunning, I.I. Fabrikant, Studies of electron-molecule scattering at microelectron volt energies using very-high-n Rydberg atoms. Phys. Rev. Lett. **75**, 810 (1995)
122. A.D. Boozer, A. Boca, R. Miller, T.E. Northup, H.J. Kimble, Cooling to the ground state of axial motion for one atom strongly coupled to an optical cavity. Phys. Rev. Lett. **97**, 083602 (2006)
123. C.-W. Chou et al., Measurement-induced entanglement for excitation stored in remote atomic ensembles. Nature **438**, 828 (2005)
124. T. Aoki et al., Observation of strong coupling between one atom and a monolithic microresonator. Nature **443**, 671 (2006)
125. B. Dayan et al., A photon turnstile dynamically regulated by one atom. Science **319**, 1062 (2008)
126. D. Felinto et al., Conditional control of the quantum states of remote atomic memories for quantum networking. Nat. Phys. **2**, 844 (2006)

Chapter 2
Rydberg Atoms and Their Interactions

Abstract This chapter provides a comprehensive overview of Rydberg atoms, with emphasis on the preparation methods, physical properties, and their extraordinary long-range interactions. Both resonant dipoledipole and van der Waals couplings are treated in detail, including the Frster resonance picture and angular dependence of interaction strength. Mechanisms for controlling interactions via electromagnetic fields and distance-selective dressing are presented, highlighting the central role of Rydberg blockade and tunability in many-body quantum dynamics and quantum information protocols.

2.1 Introduction

One of the most remarkable properties of Rydberg atoms is the strength and tunability of their mutual interactions. Due to their large principal quantum number n, Rydberg states possess enormous electric dipole moments, which scale as $n^2 e a_0$, and polarizabilities scaling as n^7. As a result, the interaction energy between two Rydberg atoms can exceed typical radiative linewidths by several orders of magnitude, even at micrometer-scale separations.

Two primary interaction mechanisms dominate the physics of Rydberg ensembles: the *resonant dipole–dipole interaction* and the *van der Waals interaction*. The resonant dipole–dipole interaction arises when two atoms are coupled through an energy-conserving exchange of excitation, leading to an interaction potential of the form

$$V_{\mathrm{dd}}(R) = \frac{C_3}{R^3}, \tag{2.1}$$

where C_3 is the dipole–dipole coefficient and R is the interatomic distance. Such resonant processes can be tuned by applying external electric or magnetic fields to bring different pair states into near-degeneracy (Förster resonances).

K. A. Yasir and W.-M. Liu, *Rydberg Atoms in Cavity*, Springer Series in Solid-State Sciences 213, https://doi.org/10.1007/978-981-95-4571-1_2

When the resonance condition is not satisfied, the interaction is dominated by off-resonant coupling, giving rise to the van der Waals potential,

$$V_{\mathrm{vdW}}(R) = \frac{C_6}{R^6}, \tag{2.2}$$

with C_6 being the van der Waals coefficient. For high-lying Rydberg states, C_6 grows rapidly with n (scaling approximately as n^{11}), making the interaction extremely strong compared to ground-state atoms.

These strong interactions underlie the *Rydberg blockade* effect, in which the excitation of one atom to a Rydberg state shifts the energy levels of nearby atoms, thereby preventing their simultaneous excitation. This phenomenon leads to collective excitation dynamics and is a cornerstone for applications in quantum information processing, quantum nonlinear optics, and the study of many-body physics in strongly interacting regimes. In this chapter, we will be discussing the properties and different types of Rydberg interactions.

2.2 Rydberg Atoms

2.2.1 Preparation of Rydberg Atoms

The usual two-level atoms, conventionally with a ground and an excited state, often have very short radiative lifetime of nanoseconds which limits their practical implementation. However, Rydberg atoms are feasible to be used in experiments because of their highly excited nature with long lifetime and have been thoroughly investigated in this regard. The concept of Rydberg atoms is connected to the Bohr's atomic model and based on the principal quantum number n which describes the electron's distance from nucleus in the form of atomic energy levels. The Rydberg atoms with quantum number $n = 50$ or more have radius thousand times larger than the ground state atoms which makes it feasible to implement effective two-level systems for quantum applications using Rydberg atomsmakes it feasible to implement effective two-level systems for quantum applications using Rydberg atoms. The inner-shell electrons—low level electrons together with nucleus form an ionic core or inner-shell - screen the charge on the nucleus of atoms, which leads towards the Rydberg atoms hydrogenic in nature. Figure 2.1 illustrates the classical and quantum pictures of Rydberg atoms. The energy spectrum of Rydberg atoms as a function of principal quantum number n and orbital angular momentum l (with total angular momentum $j = l \pm 1/2$) is given as,

$$E_{n,l} = -\frac{R_y}{(n^*)^2}, \tag{2.3}$$

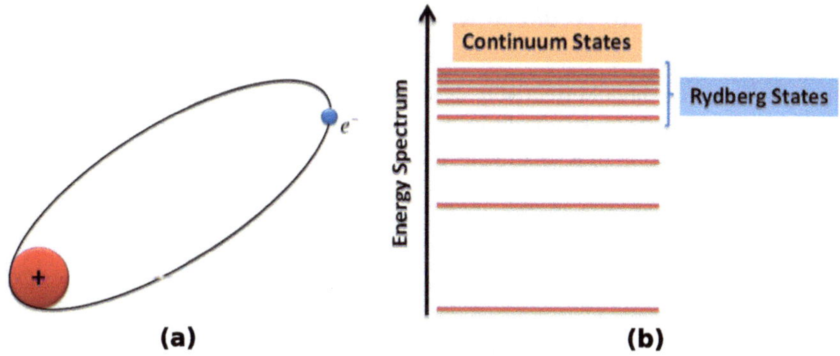

Fig. 2.1 **a** The classical picture of an alkali Rydberg atom. It is shown that the Rydberg atoms have an electron far from the nucleus which sees the inner-shell as an electronic core with resulting charge $+e$. The Rydberg electron can penetrate into the electronic core when it has a low orbital angular momentum l. Therefore the Rydberg electron should be at high angular momentum in order to stabilize Rydberg atom. **b** The quantum picture of energy spectrum of a Rydberg atom. In order to achieve Rydberg states in an atom, the energy levels should be very near to the ionization threshold. As a consequence, the spacing between energy level decreases with the increasing in principal quantum number and finally, forms the Rydberg states. For an alkali Rydberg atom, the principal quantum number is considered around $n = 50$ with the electron transition frequency between 10 and 100 GHz

where $n^* = n - \delta_l$ and $R_y \approx 13.6\text{eV}$ is the Rydberg constant. The electron with low orbital angular momentum l can penetrate inside the electronic core which increases the effective charge of inner-shell and decreases the binding energy. The parameter δ_l represents the quantum defect which becomes negligible ($\delta_l = 0$) for high angular momentum $l \geq 3$ states in alkali atoms. Rydberg atoms with energy spacing between adjacent Rydberg levels scales as $1/n^3$, leading to a quasi-continuum at large value of principal quantum number n. However, the spontaneous emission to the neighboring Rydberg states is very slow for such highly excited states of Rydberg atoms. With the help of pulsed tunable dye lasers, one can differentiate these neighboring states from each other in order to provide controlled excitations of Rydberg states to n, l and m sub-energy levels.

The most efficient method to prepare Rydberg states in atoms was proposed by D. Kleppner et al. in 1981 [1]. They used two different laser beams to transfer atoms into an intermediate state and a third laser to transform this intermediate state into Rydberg atom. One of the major obstacles is the inter-atomic collisions because these collisions break apart the Rydberg states. To deal with this obstacle, they used an atomic beam which generates sufficient spacing between atoms to prevent the collisions. They first heated alkali atoms to achieve vaporization and then sent this atomic beam to a vacuum chamber via a small hole. The vacuum inside the chamber also prevents the inter-atomic collisions. After passing through another small aperture, properly aligned Rydberg atoms were achieved. Two plates were placed below and above the atomic beam and a pulsed voltage was applied between the plates to ionize Rydberg atoms and detect freed electrons [1].

2.2.2 Properties of Rydberg Atoms

The Bohr's atomic model effectively describes the properties of almost all types of Rydberg atoms due to their similarity with hydrogen atoms. The relation between centripetal acceleration of Rydberg electron and Coulomb force exerted by the inner core can be written as,

$$\frac{m_e v^2}{r_{max,min}} = \frac{e^2}{4\pi\epsilon_0 r_{max,min}^2}, \tag{2.4}$$

where m_e is the mass of Rydberg electron with charge e and v is circular velocity of a classical electron. The maximum and minimum distance of Rydberg electron from the central core or inner-shell are classically defined as,

$$r_{max,min} = \frac{n^2 a_0(1 \pm \epsilon)}{Z}, \tag{2.5}$$

where a_0 is the Bohr radius and the parameter $\epsilon = \sqrt{1-(l+1)^2/n^2)}$ represents eccentricity of the classical electron orbit with orbital angular momentum l as defined before. This expression arises from the semiclassical Keplerian treatment of orbits and becomes valid in the $large - n,\ large - -$ regime. The closest position of electron from nucleus is known as perihelion, $r_{min} = n^2 a_0(1-\epsilon)/Z$, and the farthest position is called aphelion, $r_{max} = n^2 a_0(1+\epsilon)/Z$. The average of the electron distance represents average radius of Rydberg atoms, $r = (r_{max} + r_{min})/2 = n^2 a_0/Z$. This Rydberg radius can be defined quantum mechanically by considering quantized form of orbital angular momentum in units of Planck constant $\hbar$ as, $r = \hbar^2 n^2/(e^2/4\pi\epsilon_0)m_e = a_0 n^2$, where $m_e cr = n\hbar$. These expressions are important to understand the dependence of size of the Rydberg atoms on principal quantum number and explain how the size of the Rydberg atoms can be taken to the macroscopic level with the help of the principal quantum number. The Kepler frequency $\omega = c/r$, which is crucially important in order to discuss Rydberg atoms in cavities and in optical lattices because of the scale of n^{-3}, can be defined from above relations as, $\omega = e^2/4\pi\epsilon_0 m_e r^3$. The other important parametric scalings, under scaling law, are presented in Table 2.1.

The wave function dynamics of Rydberg states is expressed as

$$\psi_{nljm_j}(\mathbf{r}) = \langle \mathbf{r}|nljm_j\rangle = \frac{P_{nl}(r)}{r}\mathcal{Y}_{ljm_j}(\theta,\phi),$$

where $P_{nl}(r)$ is the radial wavefunction and $\mathcal{Y}_{ljm_j}(\theta,\phi)$ denotes the spin-orbit coupled angular part. These wavefunctions are conventionally calculated using the quantum defect theory, which accounts for the deviation of real atomic potentials from the ideal Coulomb potential due to the atomic core penetration and polarization effects. Quantum defect theory modifies the principal quantum number n into an effective

Table 2.1 Properties of Rydberg atoms under scaling law

Parameters	In scale of
Binding energy	n^{-2}
Energy between adjacent states	n^{-3}
Orbital radius	n^2
Dipole moment	n^2
Polarizability	n^7
Radiative lifetime	n^3
Rabi frequency for excitation	$n^{-3/2}$

quantum number $n^* = n - \delta_l$, where δ_l is the quantum defect dependent on the orbital angular momentum l. This adjustment allows the use of hydrogenic radial wavefunctions, with the quantum defect incorporated as a phase shift, and often involves special functions such as hypergeometric functions to accurately describe the radial part $P_{nl}(r)$.

An important quantity in this context is the radial matrix in dipole-relevant form (linear in r) defined as

$$\langle r \rangle_{nl}^{n'l'} = \int r \, P_{n'l'}(r) P_{nl}(r) \, dr, \tag{2.6}$$

which physically represents the expectation value or overlap of the radial coordinate r between two Rydberg states $|nl\rangle$ and $|n'l'\rangle$. These radial matrix elements play a crucial role in characterizing the interactions of Rydberg atoms with external and internal electromagnetic fields. They determine the strength of dipole transitions between states, contribute to the Stark and Zeeman shifts, and influence the long-range dipole–dipole and van der Waals interactions among Rydberg atoms. As discussed in reference [2], accurate evaluation of these elements using quantum defect wavefunctions is essential for predicting level shifts and transition probabilities.

Transition matrix elements are particularly significant for states obeying the dipole selection rule $l' = l \pm 1$, especially when the principal quantum number n is large ($n \gg 1$). At these high excitations, the Rydberg states form a nearly saturated energy spectrum with closely spaced levels, and states with different l but similar n can have very similar radial wavefunctions, increasing the overlap and hence the matrix elements. Typically, the largest radial matrix elements fall within the range $0.5n^2a_0$ to $1.5n^2a_0$, where a_0 is the Bohr radius. For example, in rubidium Rydberg atoms at around $n = 50$, Fig. 2.2a shows that the radial matrix elements connecting the $90d$ state to neighboring $91p$, $92p$, $89f$, and $88f$ states are approximately $(1.3, 0.76, 1.3, 0.80) \times n^2a_0$, respectively. This contrasts sharply with the much smaller values (less than $0.17n^2a_0$) for transitions involving other states. Therefore, even weak external electric fields can induce substantial Stark shifts among these closely spaced Rydberg levels. This sensitivity allows detailed investigations of interatomic potentials and interactions by focusing on the influence of fields on transitions

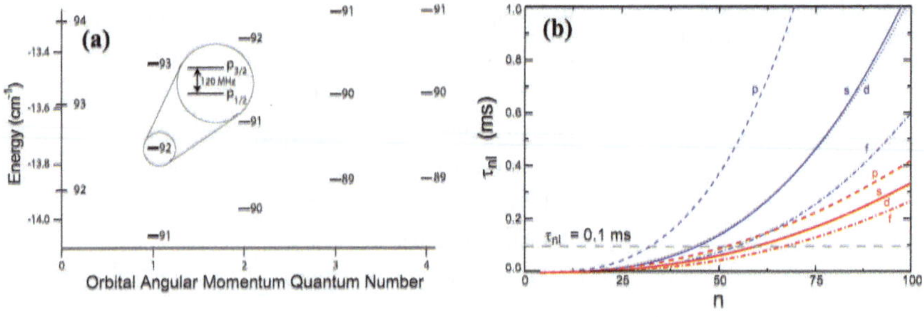

Fig. 2.2 **a** The energy level spectrum for Rydberg states of a Rb atom with principal quantum number near $n = 90$. The electron transition frequency among the Rydberg states is considered about $150MHz$. Adopted from [4], ©(2025) with permission from American Physical Society. **b** The excited state lifetime for Rb Rydberg atom as a function of principal quantum number n. The blue and red curves represent temperature $T = 0K$ and $300K$, respectively. s, p, d and f curves represent states of the Rb Rydberg atom. Obtained from [3], ©(2025) with permission from American Physical Society

between states differing by $\Delta l = \pm 1$, providing crucial insight into Rydberg atom physics and their manipulation.

Furthermore, the radiative and blackbody-limited lifetimes of Rydberg atoms with low orbital angular momentum are important to discuss here, as they directly influence the coherent dynamics of these highly excited states. The finite-temperature lifetime τ_{nl} of a Rydberg state $|nl\rangle$ can be expressed as the combined effect of spontaneous radiative decay at zero temperature and blackbody-induced transitions at finite temperature:

$$\frac{1}{\tau_{nl}} = \frac{1}{\tau_{nl}^{(0)}} + \frac{1}{\tau_{nl}^{(bb)}}, \tag{2.7}$$

where $\tau_{nl}^{(0)}$ is the zero-temperature lifetime, often modeled as

$$\tau_{nl}^{(0)} = n_l^{(0)}(n^*)^{\alpha_l},$$

with $n^* = n - \delta_l$ being the effective principal quantum number corrected by the quantum defect δ_l, and $\alpha_l \approx 3$ a constant parameter typically used for alkali atoms. The parameter $n_l^{(0)}$ is an empirical coefficient dependent on the angular momentum l.

The term $\tau_{nl}^{(bb)}$ accounts for the additional decay induced by blackbody radiation at finite temperature T. For large principal quantum numbers n, this blackbody-induced decay rate can be approximated by the formula

$$\frac{1}{\tau_{nl}^{(bb)}} \approx \frac{4\alpha^3 k_B T}{3\hbar n^2}, \tag{2.8}$$

where $\alpha \approx 1/137$ is the fine-structure constant, k_B is Boltzmann's constant, and $\hbar$ is the reduced Planck constant. This approximate expression captures the scaling behavior of blackbody-induced photoionization, which corresponds to transitions from bound Rydberg states to the continuum. Figure 2.2b shows the calculated radiative blackbody lifetimes for low angular momentum Rydberg states of rubidium atoms (s, p, d, and f states) as a function of principal quantum number n. For $n \gg 50$, the total lifetimes τ_{nl} at room temperature typically exceed 50 μs. This imposes a practical constraint on the timescales for coherent manipulation of Rydberg states, indicating that experimental protocols often need to operate on microsecond or faster (MHz-scale) excitation times to preserve high-fidelity quantum control. These characteristic lifetimes, combined with the strong, tunable interactions arising from the large dipole moments and long-range nature of Rydberg states, underpin the versatile applications of Rydberg atoms in quantum information, simulation, and nonlinear optics, as discussed in the following sections.

2.3 Rydberg Atom-Field Interaction

The radial or dipole matrix elements also imply that the Rydberg atoms are extremely sensitive to the incident electromagnetic field, which is indeed an important feature to optically manipulate and control both the strength and angular dependence in Rydberg–Rydberg interactions. This high sensitivity arises due to the large spatial extent of the electron orbitals in Rydberg states, which increases the transition dipole moment and enhances their response to external electric fields. The interaction of a small DC electric field with Rydberg atoms will generate dipole coupling of the order of $e\langle r\rangle E$, where E is the applied electric field and $\langle r\rangle$ is the expectation value of the electron's radial coordinate. Owing to the n^2 scaling of the atomic radius, this interaction can become quite significant for Rydberg atoms with high principal quantum numbers.

These dipole couplings are much less than the energy difference ΔE between neighboring opposite parity states, which consequently generates a second-order Stark effect (or Stark shift) of the order $-\alpha E^2/2$, with $\alpha \sim n^7$ being the polarizability. The large polarizability at high n values means that even weak electric fields can induce measurable energy shifts, making Rydberg atoms ideal probes for precision electrometry and tunable interaction platforms. Mostly, the Stark shifts are smaller than this upper bound because the states equal and opposite to ΔE cancel each other by producing an opposite shift, effectively creating a near-zero net perturbation in energy levels in the absence of asymmetry. This behavior is schematically shown in Fig. 2.2a, where the symmetry of the state distribution leads to partial cancellation in the energy shifts.

Despite these partial cancellations, the electromagnetic stability at the scale of $0.01(100/n)^{7/2} V/cm$ is still required to limit the Stark shift below 1 MHz, highlighting the extraordinary sensitivity of Rydberg levels to stray electric fields [4]. This stringent requirement for field control becomes particularly important in high-precision quantum optics and quantum information experiments where long coherence times are needed. Moreover, the tunability of energy levels via electric fields can also be harnessed to deliberately shift or bring Rydberg states into resonance, enabling coherent control of dipole–dipole interactions and facilitating processes such as Förster resonances, dipole blockade, and entanglement generation. Therefore, the electric field sensitivity is not only a challenge but also a key advantage in Rydberg-based quantum technology platforms.

Another feature that the electromagnetic interactions bring is the ionization of Rydberg atoms. In order to determine the ionization caused by the electromagnetic interaction with Rydberg atoms, the combined Coulomb-Stark potential is given as,

$$V_{CS} = -\frac{1}{r} + Ez, \tag{2.9}$$

where z is the direction in which the electromagnetic field E is applied and the parameter r is the orbital radius of the Rydberg atom. This definition possesses key importance to evaluate and discuss the dynamics of Rydberg atoms in cavities. To understand it further, the saddle point of the potential is computed by putting the time derivative of the potential equal to zero, which should be around $z = 1/\sqrt{E}$ along the field direction having potential energies in the order of $-2\sqrt{E}$. These relations lead us to determine the exact expression for the energy of incident electric field $E = (W/2)^2$ which ionizes Rydberg atoms, where W is the binding energy of Rydberg electrons. The binding energy $W = -1/2n^{*2}$ produces a binding force that keeps Rydberg electron associated with inner-shell and it is inversely proportional to the square of orbital radius of the Rydberg electron $r \sim n^{*2} \Rightarrow W \sim -1/r$ approximately. By using the relation for the radius of Rydberg electron orbit, we can redefine the resultant ionization energy of Rydberg atoms by using $W = -1/(2n^{*2})$, and $V(z_s) = -2\sqrt{E}$, equating as,

$$E = \frac{1}{16r^2} = \frac{1}{16n^{*4}}, \tag{2.10}$$

where n^* is the effective principal quantum number defined as, $n^* = n - \delta_l$. Here, as we know, δ_l is the parameter defining quantum defect. It should be noted that the strength of the incident electromagnetic field is very critical because it may cause atoms to face Stark shift and Rydberg states may mix with nearby energy levels. Depending on the intensity of incident light and the area of incident cross-section, the Rydberg atoms may encounter two consequences. Either they will adiabatically stay in their original energy levels or they will diabatically hop to the nearby energy states. To avoid hopping of Rydberg states to their nearby energy levels, adiabatic approximation is used which broadens the ionization spectrum by limiting the strength of the interacting electromagnetic field to $E_{Threshold} \approx 1/n^{*5}$. This adiabatic approximation is crucially important to investigate the dynamics of Rydberg atoms in cavities and has been used in almost all investigations.

2.4 Blockade Excitation

One of the most significant features that amplify the importance of Rydberg atoms is the Rydberg blockade excitation. The concept of Rydberg blockade arises from the strong interactions between Rydberg atoms that cause the shift among the energy levels of Rydberg atoms. When the atomic ensemble is excited by the incident electromagnetic field, the spectrum excited energy level appears in a different way other than the equally distant ladder representation. This is because the energy level shift in second excitation becomes so large that it makes atomic excitation frequency off-resonant with the incident laser frequency which consequently creates a blockade of all the above excited states. This shift is called the blockade shift which is denoted with B as shown in Fig. 2.3a. With the help of the blockade shift B, Rydberg blockade excitation enables us to construct a two-level system with N number of atoms having enhanced Rabi frequency $\Omega_N = \sqrt{N}\Omega$. The implementation of a single atom qubits and two (or multiple) qubit quantum logic gates are the direct applications of Rydberg blockades which are exactly in accordance to the single-atom protocols. It is also feasible with Rydberg blockade to generate higher-order excitation state $|1\rangle$ by using the modified Rabi coupling between $|0\rangle \rightarrow |1\rangle$ and $|1\rangle \rightarrow |r\rangle$, as shown in Fig. 2.3b. Such kind of highly excited states are extremely important to implement and utilize multi-photon wave-packets [4–6].

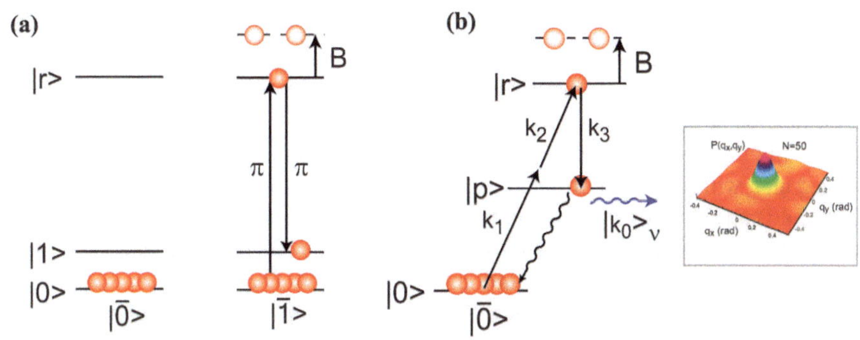

Fig. 2.3 **a** The energy level spectrum of blockade qubits (or ensemble qubits) as well as their rotational transitions between basis states $|\bar{0}\rangle$ and $|\bar{1}\rangle$. The left side shows the doubly excitation of Rydberg ensemble including $|0\rangle \rightarrow |1\rangle$ and $|1\rangle \rightarrow |r\rangle$. The right side shows the rotational transition of the qubit from $|\bar{a}\rangle$ to blockaded state with a π Rabi cycle. Here B is the two-atom blockade shift produced by the double excitation. From [4], ©(2025) with permission from American Physical Society. **b** The schematic diagram for the preparation of singly excited state, $|0\rangle \rightarrow |p\rangle$ and $|p\rangle \rightarrow |r\rangle$, which will decay after emitting a single photon having phase shift along $\mathbf{k}_0$. The wave-vector $\mathbf{k}_1$ and $\mathbf{k}_2$ correspond to the single excitation from ground state to the state $|r\rangle$ while $\mathbf{k}_3$ is for decay of the qubit to the intermediate state $|p\rangle$ by emitting a photon with state $|k_0\rangle$. The inset figure illustrates the probability distribution this state $P(q_x, q_y)$ as a function of position quadrature of qubit in x and y directions. Adopted from [4], ©(2025) with permission from American Physical Society and [8], ©(2025) with permission from American Physical Society

The blockade effects also cause the double excitations of states which are off-resonant with the excited state shifted by the factor B. The probability of these double excited states can be defined as [4],

$$P_2 = \frac{N-1}{N}(\frac{\Omega_N^2}{2B^2}). \tag{2.11}$$

It can be noted that the probability of having unwanted states is inversely proportional to the blockade shift B. $\Omega_n = \sqrt{N}\Omega$ is the collective Rabi oscillation frequency of Rydberg ensemble. It means that for the higher values of blockade shift, there are fewer chances to have such kind of double excitations. The blockade shift in this regard is defined as [7, 8],

$$\frac{1}{B^2} = \frac{2}{N(N-1)}\sum_{\varphi}\sum_{i<j}\frac{|k_{\varphi ij}|^2}{\Delta_{\varphi ij}^2}, \tag{2.12}$$

where $k_{\varphi ij} = \langle\varphi|H_{\text{int}}|ij\rangle$ and $|\varphi\rangle$ represents the excited state of Rydberg atom i and j and $\Delta_{\varphi ij}$ corresponds to the dipole–dipole energy shift among the excited states of the atoms. The probability of double excited states depends on the Rabi coupling associated to the pair of excited states $|\varphi\rangle$ and on the dipole–dipole energy shift $\Delta_{\varphi ij}$ among the states. Therefore, the blockade shift B is scaled with the dimensionless parameter $k_{\varphi ij}'$. It is clear from Eq. (2.12) that the small amount of atom-atom interactions in Rydberg ensemble can modify the value of blockade effects and decrease the width of the shift which may lead to the situation where the chances for the emergence of double excited states are maximum. To realize this situation, if we model Rydberg ensemble as an electronic circuit then the blockade shift B will be the impedance of the circuit formed by the parallel combination of components having individual impedance with the strength about $\approx (N^2/2)[\sum_{i<j}|k_{\varphi ij}|^2/\Delta_{\varphi ij}^2]^{-1}$ [4]. The weak blockade can only allow the excitation of a pair of atomic states $|\varphi\rangle$ and it cannot trigger Rydberg blockade excitations. These kind of effects limit the visualization of the strong Rydberg blockade excitations.

The challenge to achieve strong Rydberg blockade excitation has encouraged researchers to further understand the many-body interactions among the atoms and their excited states [9]. These motivations led researchers to transfer interactions between coupled Rydberg ensemble form long-range $1/r^3$ Förster interactions [10] to a short-range $1/r^6$ van der Waals interactions [11, 12], as discussed in the following chapters. Depending upon the special distribution and orientation of atomic states, it was also found that the Rydberg atoms can be coupled to non-interacting Förster zero states, which may facilitate the manipulation of Rydberg blockades [7]. Further, the addition of another atom to the two-atom Rydberg state provides an opportunity to generate non-interacting zero states by manipulating three-photon Rydberg excitations, even under the environment where the previous two-photon excitations are strongly blockaded [13]. The powerful concept of Rydberg blockade lies in the foundation of many important and crucial scientific achievements, such as, the realization of many-body electromagnetically induce transparency [14], entanglement via Rydberg blockade [11] and achieving simulation of dissipative quantum many-body systems [15].

Aside from the advantages of Rydberg blockades in light-matter interface Rydberg blockade also possesses key importance in achieving and sustaining controlled emission spectrum to observe symmetric and long distance entanglement in Rydberg states. Figure 2.3b illustrates the preparation of blockade state of Rydberg ensemble which is excited through multi-photon process and can be defined as,

$$|\Psi_0\rangle = \frac{1}{\sqrt{N}} \sum_{j=1}^{N} e^{ik_0.r_j} |p_j\rangle \otimes |0\rangle_v, \tag{2.13}$$

where $|p_j\rangle$ describes the Rydberg state with atom j and parameter r_j corresponds to the position of excited atoms. While state $|0\rangle_v$ represents the vacuum state created by the incident photon field. The state for the Rydberg ensemble, sharing single excitation with complex phase, is engineered by using tree-laser configuration. When first and second lasers with wave-vector k_1 and k_2, respectively, interact with Rydberg atomic ensemble, they will generate Rydberg excitation blockade that will limit number of atoms to one by blocking the excitation of all other atoms in ensemble. To avoid this effect, another laser, with *pi* configuration and wave-vector k_3, is used which excites Rydberg atoms to the state $|p\rangle$, as described in Eq. (2.13), where the resultant wave-vector is defined in the combination form of incident wave-vectors $k_0 = k_1 + k_2 + k_3$. The spontaneous decay of the emerging Rydberg state will convert it into form $|\Psi\rangle = |\bar{0}\rangle \otimes |k_0\rangle_v$ by transferring all atoms to their respective ground states. Here the photon, emitted in this process, will be in-phase to incident wave-vector k_0. For this purpose, the Rydberg-Rydberg interaction in atomic ensemble should be stronger than the excitation frequency of the atoms. Here it should also be noted that the size of blockade should be larger than the size of atomic ensemble itself otherwise multiple blockade may excite. The process of creating such Rydberg states magnifies their application towards coherent photons sources and lasers. In the following sections, we will discuss atom-atom interactions in Rydberg ensembles.

2.5 Dipole–Dipole Interactions

The integration of Rydberg atoms in cavities to perform collective quantum electrodynamics is significantly useful due to the sensitive nature of Rydberg atoms to electromagnetic fields. This integration enables the realization of hybrid light–matter platforms in which the long-range interactions of Rydberg atoms combine with the enhanced photon exchange mediated by high-finesse cavities, opening up new possibilities for engineering nonlinear quantum optics and strongly correlated quantum many-body systems. The collective dynamics are only possible in the presence of strong inter-atomic potential energies arising from the strong dipole–dipole interaction between Rydberg atoms, which scales rapidly with the principal quantum number and can dominate over other forms of decoherence or thermal effects in ultra-cold atomic ensembles.

As stated previously, the strong Rydberg blockade excitation is only possible when the Rydberg–Rydberg interactions are stronger than the inter-atomic Rabi coupling. This blockade mechanism ensures that the excitation of a single atom to a Rydberg state inhibits the simultaneous excitation of its neighbors within a characteristic blockade radius, thereby allowing for deterministic preparation of collective quantum states. Under this condition, the first-order Rydberg–Rydberg interaction will be suppressed and become ineffective to the entangled properties of Rydberg atoms, since no two atoms can be simultaneously excited within the blockade volume. This leads to the emergence of highly nonlocal and collective behaviors in the atomic ensemble, such as the formation of so-called "superatoms," where an ensemble of atoms collectively behaves as a single effective two-level system with enhanced coupling strength proportional to $\sqrt{N}$, where N is the number of atoms within the blockade region.

Such a situation provides a precise control over the low-temperature dynamics and spatial distribution of Rydberg atoms, making them usable in versatile configurations, including quantum information processing, photonic switching, and quantum simulations of spin models and lattice gauge theories. In particular, the strong interactions combined with cavity-mediated photon fields enable the generation of long-lived entangled states and collective polaritonic modes with adjustable interaction strength and coherence properties. These features are instrumental in designing quantum interfaces, few-photon transistors, and entangled photon sources in cavity QED platforms incorporating Rydberg ensembles.

In the following sections, we will further discuss and quantify the strength of such inter-atomic interactions in Rydberg atoms, particularly focusing on their scaling laws, angular dependence, and tunability via external electric and microwave fields. Moreover, we will analyze how these interactions can be harnessed in cavity setups to realize deterministic photon-photon gates, collective dark states, and hybrid spin-photon entanglement protocols essential for future quantum technologies.

If we consider two Rydberg atoms (**A** and **B**) separated by the distance **R**, as shown in Fig. 2.4a, with same initial states $|\Psi_0\rangle = |n, l, j, m_j\rangle$ under the condition that the distance between the atoms should be greater than the orbital radius of the Rydberg electron ($\mathbf{R} >> a_0 n^2$), then the leading energies of dipole–dipole interaction can be written as,

$$V_{dd}(\mathbf{R}) = \frac{\mu_1 \cdot \mu_2}{R^3} - \frac{3(\mu_1 \cdot \mathbf{R})(\mu_2 \cdot \mathbf{R})}{R^5}, \tag{2.14}$$

where μ_1 and μ_2 account for the dipole moment of the atomic transition from state Ψ_0 to Ψ_1 and Ψ_2, respectively. If we consider the distance vector **R** aligns along

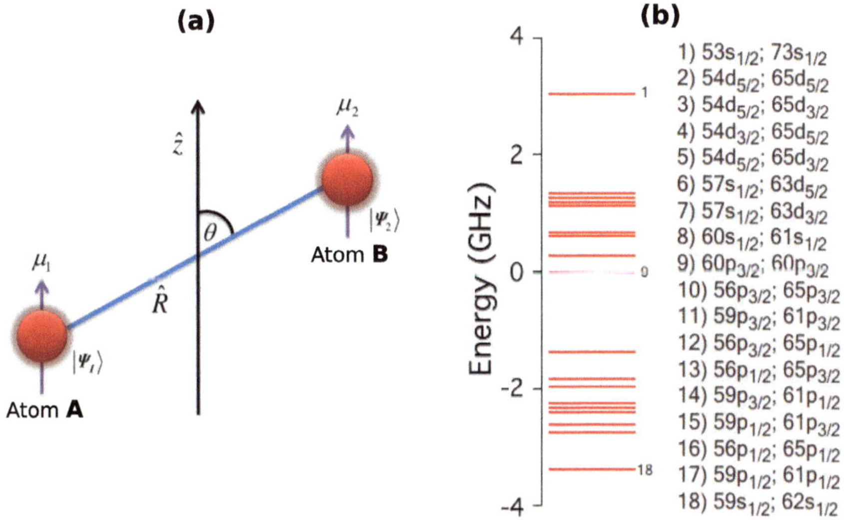

Fig. 2.4 **a** The systematic picture of dipole–dipole interaction between atoms **A** and **B** separated by the distance R, which makes angle θ with $\hat{z}$-axis. μ_1 and μ_2 are the electric dipole moment for atoms **A** and **B**, respectively. **b** The combination of near neighboring pairs of states connected to the central state $|60p_{3/2}60p_{3/2}\rangle$ under dipole–dipole interactions. Adopted from [4], ©(2025) with permission from American Physical Society

$\hat{z}$-axis, as shown in the Fig. 2.4a, which is only possible when the inclined angle of the vector $\mathbf{R}$ is zero ($\theta = 0$), the dipole–dipole energy can be reduced to the form as,

$$V_{dd}(\mathbf{R}) = \frac{\mu_{1-}\mu_{2+} + \mu_{1+}\mu_{2-} - 2\mu_{1z}\mu_{2z}}{R^3}, \tag{2.15}$$

where the subscripts (,) represent the polarization components of the atomic dipole moments along the spherical basis directions. This decomposition reveals the contributions from different angular momentum exchange channels: $^+$, $^+$, and ,. These interaction terms conserve the projection of angular momentum along the interatomic axis and are especially useful for quantifying state-specific couplings in Rydberg ensembles. To calculate energy shift in the level spectrum caused by the dipole–dipole interaction, it is easy to transform and use the pair states instead of individual atomic states. These pair states have resultant angular momentum in the combinational form of individual angular momentum of atomic states $j_T = j_1 + j_2$ [4].

To further analyze dipole–dipole interaction and the behavior of these paired atomic states, which are crucially important to understand the concept of Rydberg blockades, we define the two-atom state coupled to the incident laser field and with extremely large orbital radius of Rydberg electron $R \approx \infty$ as,

$$|\Psi_2\rangle = |nlj\rangle_A \bigotimes |nlj\rangle_B. \tag{2.16}$$

The influence of dipole–dipole interaction on such kind of state possesses a fundamental value in order to create strong Rydberg blockade. However, there are some particular configurations where by keeping l and j fix and changing the value of principal quantum number n, we can have such strong dipole–dipole interactions but it's difficult to illustrate experimentally. If we switch-off the incident laser field, the two-atom state still have degeneracy of the order of $(2j_T + 1)^2$, where j_T is the combined electron angular momentum of both atoms. The dipole–dipole interaction not only alter the two-atom state but also trigger its transition to the neighboring two-atom states, where the dipole selection conditions are satisfied, as stated above. In fact not only neighboring but an infinite number of surrounding states, including continuum states of the Rydberg atom, can also get excited due to the dipole–dipole interactions. However, conventionally, dipole–dipole interactions for the nearby states having the same characteristics are considered, as illustrated in Fig. 2.4b, where the energy level spectrum, around state $|60p_{3/2}60p_{3/2}\rangle$, for the Rb Rydberg atoms is shown. It can be noted that, for the fixed large quantum number $g = \pm 8$, the multiple pairs of coupled two-atomic states between the frequency range of -4GHz to $+4$GHz. Interestingly, even in presence of large number of other coupled pair of states, coupling between state $|60p_{3/2}60p_{3/2}\rangle$ and nearby state $|60s_{1/2}60s_{1/2}\rangle$ generates strong dipole–dipole interactions. It is understandable that the small energy level gap and large matrix elements are the basic reason behind these effects. As the other neighboring states have small matrix elements and large energy level gap, therefore, they do not play a significant role in the excitation of dipole–dipole interaction. However, combination of states near -2GHz shows some effects on the interactions produced by $|60p_{3/2}60p_{3/2}\rangle$ but they can be ignored [7]. Thus, the strong and long-range dipole–dipole interactions are dominated by the central pair of states $|60p_{3/2}60p_{3/2}\rangle$ and $|60s60s\rangle$.

Generally saying, the long-range dipole interactions between coupled Rydberg ensembles are the result of the coupling between these states, $nlj + nlj$ and $n_a l_a j_a + n_b l_b j_b$ with an energy defect parameter $\delta = E(n_a l_a j_a) + E(+n_b l_b j_b) - 2E(nlj)$, where E is for the energy of the corresponding state. Another thing that should be noted that as we are interested in Rydberg blockade dynamics therefore, we are not mentioning the actual population transfer among these coupled Rydberg states. The strength of Rydberg-Rydberg interactions is further characterized by the Förster interaction picture and van der Waals interaction picture.

2.5.1 Förster and Van der Waals Picture

In Förster interaction picture, the two-level approximation on Rydberg atoms is implemented through Förster eigenstates which is fundamentally a linear combination of states from nearby channel. By representing wave function for the channel $nlj + nlj$ as $|\varphi\rangle$ and the wave function for nearby coupled states from channel $n_a l_a j_a + n_b l_b j_b$ as $|\mathcal{X}\rangle$, the Förster dipole–dipole interactions in the form of time-independent Schrödinger equation can be written as,

$$\begin{pmatrix} \delta & V_{dd} \\ V_{dd}^{\dagger} & 0 \end{pmatrix} \begin{pmatrix} |\mathcal{X}\rangle \\ |\varphi\rangle \end{pmatrix} = \Delta \begin{pmatrix} |\mathcal{X}\rangle \\ |\varphi\rangle \end{pmatrix}, \tag{2.17}$$

where V_{dd} is the dipole–dipole interaction matrix coupling the pair states $|\mathcal{X}\rangle$ and $|\varphi\rangle$, with elements scaling as $\propto C_3/R^3$, where C_3 depends on radial and angular dipole matrix elements with $\mathcal{O} \in \{1, 2\}$. If the parameter $\mathcal{O} = 1$ then the both neighboring channel will have the same strength, $(n_a l_a j_a) = (n_b l_b j_b)$ and if $\mathcal{O} = 2$ then the both channels will not be the same. The terms I_x and I_φ are the identity matrix corresponding to the wave functions $|\mathcal{X}'\rangle$ and $|\varphi\rangle$, with dimensionless Hilbert space components $(2j_a + 1)(2j_b + 1)$ and $(2j + 1)^2$, respectively. The parameter Δ represents energy eigenvalues in the Förster interaction picture. Here the wave function has the key importance to implement Förster interaction picture and can be written as,

$$|\mathcal{X}\rangle = \frac{V_{dd}}{\Delta - \delta}|\varphi\rangle. \tag{2.18}$$

It can be seen that the dynamics of the neighboring states are directly proportional to the primary wave function depending upon the strength of dipole–dipole interactions among them and quantum defect δ. By using this relation in the first line of the Eq. (2.17), we can derive the expression for the nonlinear eigenvalues of wave function $|\varphi\rangle$ [7],

$$\frac{V_{dd}^{\dagger} V_{dd}}{\Delta - \delta}|\varphi\rangle = |\varphi\rangle. \tag{2.19}$$

Note that Eq. (2.19) is a nonlinear eigenvalue equation in Δ. The solutions of this equation determine the eigenvalues and eigenvectors for the wave function $|\varphi\rangle$ depending upon the elements of the matrix $V_{dd}^{\dagger} V_{dd}$. As the matrix elements of $V_{dd}^{\dagger} V_{dd}$ have orbital radius of Rydberg electron in order of $1/R^6$ and have common characteristics with elements $C_3 = e^2 \langle r \rangle_{nl}^{n_a l_a} \langle r \rangle_{nl}^{n_b l_b}$, therefore, the eigenstates for the van der Waals picture can be directly obtained from $V_{dd}^{\dagger} V_{dd}$. The eigenvector $\mathcal{D}_\varphi$, which mostly ranges from 0 to 1, for these van der Waals eigenstates can be defined as with matrix elements,

$$V_{dd}^{\dagger} V_{dd}|\varphi\rangle = \frac{C_3^2}{R^6}\mathcal{D}_\varphi|\varphi\rangle. \tag{2.20}$$

where $\mathcal{D}_\varphi$ is a dimensionless eigenvalue encoding angular dependence of the pair state. By utilizing this expression and Eq. (2.19), one can develop relation defining Förster energy eigenvalues Δ,

$$\Delta_\varphi(R) = \frac{\delta}{2} - sgn(\delta)\sqrt{\frac{\delta^2}{4} + \frac{C_3^2}{R^6}\mathcal{D}_\varphi}, \tag{2.21}$$

where sgn(δ) accounts for positive or negative Förster energy defects due to the quantum defect shift. This relation possesses a significant value in order to develop and understand the relation between Förster eigenstates and van der Waals eigenstates. Further, this relation constitutes a major part in developing potential curves for Rydberg states as a function of R, especially for the central $nlj + nlj$ eigenstates at higher values of R. As these states are responsible for the energy shift emerging from Rydberg-Rydberg interaction, after interacting with external electromagnetic field, therefore, it is necessary to understand them in order to study Rydberg atoms in cavity environment. The quantum defect energy term can be redefined by following the above mentioned relation as, $\Delta_\varphi \approx -C_3\sqrt{\mathcal{D}_\varphi}/R^3$. Here R_c is the crossover inter-atomic distance, which is also known as critical distance because it has crucial value to determine the strength of interaction among Rydberg atoms. Furthermore, it also describes the transition of van der Waals picture to the resonant forms. When $R >> R_c$, the energy shift is defined by the van der Waals picture in the form $\Delta_\varphi \approx C_3^2 D_\varphi/\delta R^6$. On the other hand, when $R << R_C$, the two channel lie in degenerate regime where the energy shift can be defined as, $\Delta_\varphi \approx -sgn(\delta)C_3\sqrt{\mathcal{D}_\varphi}/R^3$. This is the highest value of dipole–dipole interaction between Rydberg ensembles. Figure 2.5a demonstrates the critical effects of R_c on the 60s Rb atoms with energy defect $\delta = 1.7$GHz along with the energy shift at $\delta = 0$ resulting because of the reduction in the interaction energy of the original state. Such spatial dependence of dipole–dipole interaction can be categorized, depending upon the critical value of inter-atomic distance R_c, into long-range $R >> R_c$ and short range $R << R_c$ dipole–dipole interactions. In order to investigate quantum electrodynamics of

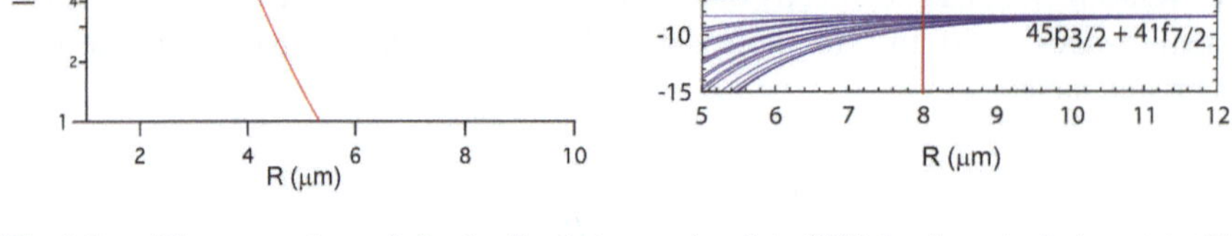

Fig. 2.5 **a** The comparison of dipole–dipole interaction ($|\Delta_\phi(R)|$) for the actual eigenstate 60s of the Rb Rydberg atoms with Förster eigenstate, as a function of inter-atomic distance R. For Rb atomic states, the quantum defects are considered non-zero $\delta \neq 0$ while for the Förster case $\delta = 0$. One can observe the reduction in dipole–dipole interaction with quantum defects. **b** The energy distribution, as a function of inter-atomic distance, for the pairs of $43d_{5/2} + 43d_{5/2}, 45p_{3/2} + 41f_{5/2}$ and $45p_{3/2} + 41f_{7/2}$ Rb atomic states. R_c is the critical value of inter-atomic distance which distinguish dipole–dipole interaction from van de Waals interactions. These results are adopted from [4], ©(2025) with permission from American Physical Society

Rydberg atom in optical cavities, the dipole–dipole interaction should be in long-range regime, where we can get strongly interacting Rydberg blockades with van der Waals picture [7].

The van der Waals picture provides an opportunity to scale principal quantum number in dipole–dipole interaction among Rydberg atoms to the order of n^{11}. However, the quantum defect moves Förster defect value near to zero which consequently limits the value of principal quantum number. To implement and utilize Rydberg atoms and blockades, it's necessary to operate them in the regime where the principal quantum number has higher values. Which is not possible in Förster interaction picture where we can only scale principal quantum number to n^4. Therefore, it's better to operate dipole–dipole interaction in Rydberg ensembles in van der Waal picture [7, 16, 17].

If we switch off the external laser field, there are limited numbers of van der Waals eigenvalues depending upon the set of angular momentum. Its is because of of Zeeman shift between energy level spectrum of the of the atoms which trigger degeneracy process for the states. Even at $\mathcal{D}_\varphi = 0$, one or more than one Förster zero states may exist depending upon the configuration of the channel of the states [7]. Such Förster-zero states arise due to angular momentum selection rules leading to $\mathcal{D}_\varphi \approx 0$, where the dipole–dipole interaction vanishes. It means such Förster zero states exist in almost all Rydberg atomic channels except where the condition $j_a = j_b = j + 1$ is imposed, means that the neighboring states are resonant with the initial state. Not only that, these Förster zero states can further get excited in presence of external electromagnetic field which results in the form of doubly excited states and these doubly excited states prevents the formation of Rydberg blockades. Even if we couple a number of channels through dipole–dipole interaction, there are still one or two states where the van der Waals interaction have zero magnitude. However, the example of such system which can promise the excitation of Rydberg blockade is illustrated in Fig. 2.5b. Figure 2.5b demonstrates channel of states of Rb atoms, with interactions $43d_{5/2} + 43d_{5/2} \rightarrow 45p_{3/2} + 41f_{5/2,7/2}$, having extremely low quantum defect $\delta = -6.0, -8.3$MHz corresponding to the angular momentum $j_{41f} = 5/2, 7/2$, respectively [7, 18]. Such kind of configuration provides a suitable environment to excite Rydberg blockades. The combination of the these states, especially $41f$ state with angular momentum $5/2, 7/2$, breaks the condition for the existence of Förster zero states. However, there are still chances for poor blockade excitation for the certain pair of states. In Fig. 2.5b a pair of states $43d_{5/2} + 43d_{5/2}$ with large values of R shows such combination. To generalize the condition for strongly interacting Rydberg blockades, it should be noted that the s states usually do not have Förster zero states because of splitting of fine-structure into the p level where Förster zero states are forced to degenerate with $\mathcal{D}_\varphi = 4/3$. But these states are limited by the experiments because the transitional Rabi frequency below the p level approaches to the factor 3 which is less the transitional frequency of level d. Therefore, these problems bond researchers to obtain blockade excitation by remaining in level d. However, a number of recent experiments have shown advancements in this regard [19, 20].

2.6 Angular Dependence

We have discussed that the characteristics of Rydberg blockade strictly depends on the strength of the dipole–dipole interactions determine by the blockade shift. In absence of the external electromagnetic field, the rotational dynamics of Förster states, which intuitively defines blockade shift, only depends on the distance R_{ij} between two neighboring atoms i and j. But in presence of external electromagnetic field, the blockade shift not only depends on the inter-atomic distance but also depends on the coupling between corresponding eigenstates and incident laser field $\kappa_{\varphi ij}$. The coupling strength $\kappa_{\varphi ij}$ is determined by measuring angle between the axis along with inter-atomic distance R_{ij} is oriented and the polarization axis of the incident electromagnetic field. If we consider an ensemble of atoms distributed in three-dimensional space (or a spherical distribution) and we shine electromagnetic field in a similar way, the incident laser field will be equally coupled to all the atomic pair states, including weakly interacting pairs, because of having similar angle to all inter-atomic orientations. But in the case of cavity, the situation is different. As in cavity, atoms are distributed along the cavity axis, therefore, the blockade shift will crucially depend on the direction of incident laser field, as shown in cavity configuration presented in Fig. fig1. Further, the blockade shift is connected to the dipole–dipole interaction strength in the form of inversely square relation, which makes blockade shift particularly dependent on small Förster eigenenergies. In other geometrical shapes of atomic orientation, there are maximum chances that the effects of small Förster eigenenergies can get enhanced by the polarization in other dimensions which will consequently decreases the coupling between incident laser and the atomic pair state $\kappa_{\varphi ij}$. However, in spherical distribution, these effects can be minimized and strong blockade excitation can be implemented as discussed in Ref. [7].

The angular dependence plays an important role and is also very crucial in order to investigate Rydberg blockade in cavity environments, as shown in Fig. 2.6a, where the behavior of blockade shift have been shown when Rydberg states are excited by the incident electromagnetic field with polarization along $\hat{z}$-axis. One can see that, for the pair of atoms with s states, the Rydberg blockade shape is nearly spherical because s states have fine structure splitting towards the p level states. However, for atomic pair states $43d_{5/2}$, there are significant variation in angle description which will strongly enhance the Förster defects for $43d_{5/2}$, leading to the anisotropic interactions. Therefore, to limit this anisotropic behavior, it's compulsory to operate with the pair of atomic states equal or near to $55s_{1/2}$. To implement van de Waals picture, we assume a state of Rydberg atoms with same angular momentum l and having the dipole–dipole interactions remain in first order which reveals the inter-atomic distance in the order of $1/R^3$. The van de Waals interaction will be suppressed because of having long-range characteristics, as shown in Fig. 2.6b. The implementation of van de Waals interactions provides an opportunity to engineer long distance entanglements and quantum gates, especially in cavities. Figure 2.6b provides a comparison of $s + p_{3/2}$ atomic states to the van de Waals interaction picture and demonstrates the quantity of symmetry-breaking as function of the angular dependence.

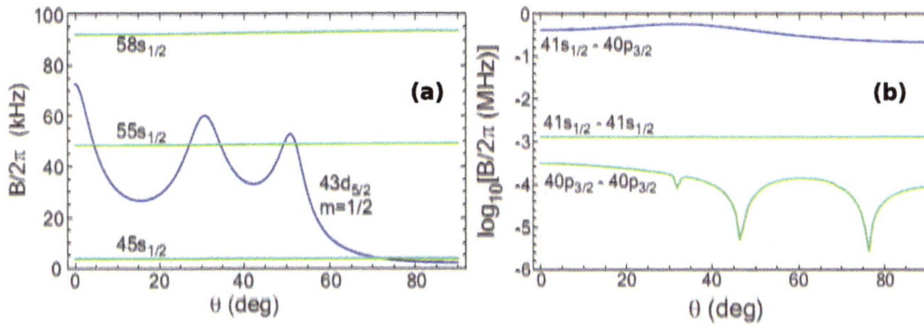

Fig. 2.6 **a** The blockade shift $B/2\pi$ as a function of angle θ between inter-atomic distance axis and polarization axis of the incident laser field, which is considered as $\hat{z}$-axis. It shows the angular dependence of states $43d_{5/2}$ (with magnetic quantum number $m = 1/2$), $45s_{1/2}$, $55s_{1/2}$ and $58s_{1/2}$ for inter-atomic distance $R = 10\mu m$. **b** The logarithmic blockade shift $log[B/2\pi]$ as a function of angle θ in order to provide comparison between interacting pairs of $41s_{1/2} - 40p_{3/2}$, $41s_{1/2} - 41s_{1/2}$ and $40s_{3/2} - 40p_{3/2}$ Rb atoms and van de Waals blockade picture for the same pairs of atoms. The magnetic quantum number is fixed to $m = 1/2$ for all results. Adopted from [4], ©(2025) with permission from American Physical Society

2.7 Electromagnetically Controlled Rydberg Interactions

The coupling of external electromagnetic field with Rydberg atomic pair states provides an indirect tool to control the strength of interaction in Rydberg ensembles, which motivated researchers to engineer Rydberg atoms in cavities. If the electromagnetic field, strong enough to mix the atomic states with nearby levels, interacts with set of pair stats of Rydberg atoms, it will limit dipole–dipole interaction by generating electric dipole moment. At this point, the strength of dipole–dipole interactions as a function of electric dipole moment can be defined as,

$$V_{dd} = \frac{1}{R^3}\left(\langle \mathbf{d} \rangle \cdot \langle \mathbf{d} \rangle - 3(\langle \mathbf{d} \rangle \cdot \hat{\mathbf{R}})(\hat{\mathbf{R}} \cdot \langle \mathbf{d} \rangle)\right), \tag{2.22}$$

where parameter $\langle \mathbf{d} \rangle$ defines the quantity of the electric dipole moment which is in order of $n^2 e a_0$, where a_0 is the Bohr's radius. The electric dipole moment increases the strength of dipole–dipole interaction to a level, where long-range blockade are possible even in presence of anisotropic interactions. The only thing that makes problem is the broadening of energy spectrum lines of Rydberg ensembles which consequently increases the quantum mechanical interference between wave function of eigenstates. However, a lot of work has been done in order to overcome and even utilize these perturbed situation in Rydberg atom optics. The incident electromagnetic field can also be used to modify Zeeman shift between atomic energy levels which will lead to the transformation of these states into Förster resonant states. This will shift the order to dipole–dipole interaction to $n^4 e^2 a_0^2 / R^3$. Such kind of behavior in Rydberg blockade under interaction of strong laser field was first studied in Ref. [21]

and verified in Refs. [22, 23]. This process can generate strongly interacting and long-range Rydberg blockades but still anisotropic interactions exist limiting their efficiency.

However, this problem has been addressed in many recent investigations [24, 25]. We can also minimize the broadening of energy levels by making excitation of the states 32d independent to the incident electromagnetic field, which is possible by coupling optical mode to the Förster resonant states, as studied in Ref. [26] and shown in Fig. 2.7a. They also develop the relation for angular dependence of dipole–dipole interaction by arranging ensemble of atoms in the form of array. If we make incident electromagnetic field resonant with microwave range, it will further modify the dynamics of Rydberg atoms by changing the strength of Förster eigenstates, as studied in Ref. [27], where they use an AC-Stark shift to modify blockade dynamics by making incident field resonant with the Förster eigenstates.

Furthermore, the interaction of Rydberg atoms with external electromagnetic field can also minimize the creation of Förster zeros by similarly modifying Zeeman shift between eigenenergy spectrum, as shown in Fig. 2.7b, where the blockade shift of the states $43d_{5/2} + 43d_{5/2}$ is plotted against applied magnetic field. The states with maxi-

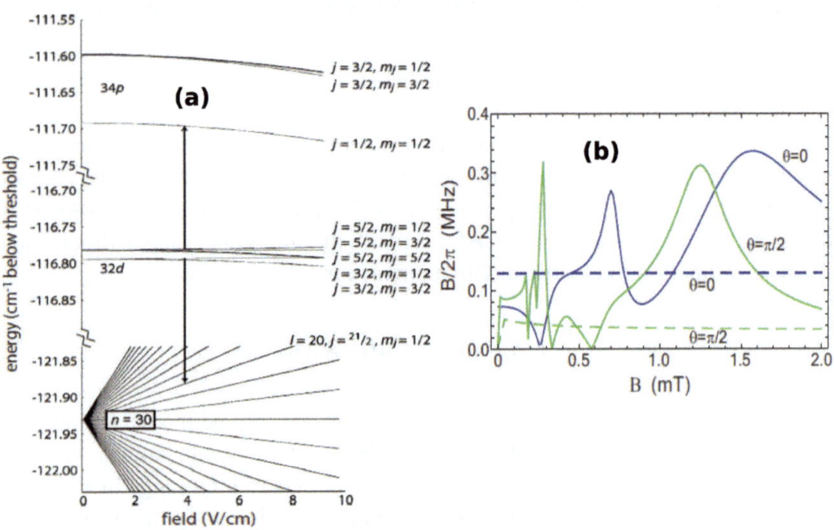

Fig. 2.7 **a** The energy spectrum as a function of incident electromagnetic field for atomic states 34p and 32d along with different values of quantum numbers j and m to demonstrate the controllability of Rydberg interaction with incident field. Lower panel shows the energy spectrum of Rydberg states by considering principal quantum number $n = 50$. Adopted from [26], ©(2025) with permission from American Physical Society. **b** The blockade shift $B/2\pi$ for $43d_{5/2}$ Rb atomic states as a function of incident magnetic field at inter-atomic distance $R = 10\mu m$ and for different incident field angles θ, as mentioned in plot. Here the solid line represents magnetic quantum number $m = 1/2$ while the dashed line is for $m = 5/2$. From [4], ©(2025) with permission from American Physical Society

mum magnetic quantum number $m = 5/2$ will be unaffected because of less Zeeman field effects. But states with magnetic quantum number $m = 1/2$ are strongly dependent on the applied magnetic field because it breaks the Zeeman degeneracy states into pair of states. Figure 2.7b also illustrates the angular dependence of blockade shift in Rydberg ensembles for different value of angle between incident laser field polarization axis and inter-atomic distance orientation.

2.8 Distance-Selective Interactions in Dressed Rydberg Atoms

The groundbreaking work by Levine et al. [28] establishes a novel paradigm for controlling quantum interactions in neutral atom arrays through the innovative use of Rydberg dressing techniques. At the heart of this achievement lies the precise engineering of distance-dependent interactions in a one-dimensional optical lattice containing ultracold rubidium atoms cooled to temperatures below 10 μK. The experimental protocol begins with the careful preparation of atoms in their electronic ground state $|g\rangle = \left|5S_{1/2}, F = 2\right\rangle$, which are then coupled to a high-lying Rydberg state $|r\rangle = \left|nS_{1/2}\right\rangle$ (with principal quantum number n ~60–80) via a two-photon transition mediated by a 780 nm and 480 nm laser system. The detuning Δ of this transition, typically ranging from $2\pi \times 5$ to 20 MHz, plays a crucial role in determining the effective interaction characteristics.

The dressed-state picture reveals the emergence of an extraordinary interaction potential $V_{\text{eff}}(R) = V_0/[1 + (R/R_c)^6] + \hbar\Delta$, where the critical blockade radius R_c exhibits a strong dependence on both the Rabi frequency Ω (typically $2\pi \times 1 - 5$ MHz) and detuning through the relation $R_c \propto (\Omega/\Delta)^{1/6}$. This potential creates a remarkable situation where atoms separated by less than $R_c \approx 5 - 10\,\mu$m experience strong repulsive interactions that prevent simultaneous Rydberg excitation, while those at larger distances interact only weakly. The experimental verification of this effect, shown in Fig. 2.8a–c, employs a sophisticated detection scheme combining fluorescence imaging with single-atom resolved microscopy, achieving spatial resolution better than 0.5 μm [29].

The quantum control capabilities of this system become particularly evident in Fig. 2.8d–f, where the researchers demonstrate coherent manipulation of specific atom pairs. By carefully tuning the interaction parameters to create a $\sqrt{\text{SWAP}}$ gate operation between atoms separated by precisely R_c, they achieve state transfer fidelities exceeding 90%, as verified through quantum process tomography. The fidelity is primarily limited by several technical factors: laser phase noise contributing ~3% error, atomic position fluctuations of about 0.1 μm (~2% error), and spontaneous emission from the intermediate state during Rydberg excitation (~1% error). Remarkably, the system maintains its performance even with up to 10% variation in atomic spacing, demonstrating exceptional robustness against positional disorder.

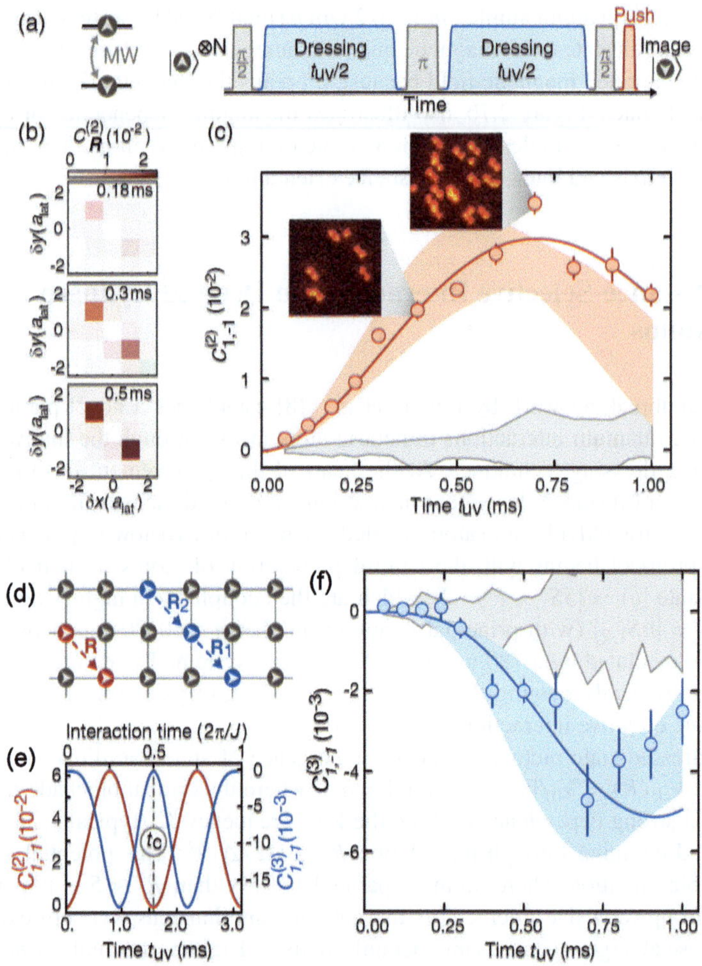

Fig. 2.8 Demonstration of distance-dependent Rydberg interactions. **a** Experimental sequence showing off-resonant dressing lasers ($\lambda = 480\,\text{nm}$, $\Delta/2\pi = 10\,\text{MHz}$, $\Omega/2\pi = 3\,\text{MHz}$) creating dressed states with tunable interactions. **b** Measured Rydberg excitation probability versus interatomic distance R, showing the characteristic blockade suppression ($P_{\text{exc}} < 0.1$) for $R < R_c \approx 7.5\,\mu\text{m}$ and the transition to uncorrelated excitation ($P_{\text{exc}} \approx 0.25$) at $R > 2R_c$. **c** Interaction potential extracted from spectroscopic measurements (dots) with theoretical fit (line) to $V_{\text{eff}}(R) = C_6/(R^6 + R_c^6)$, yielding $C_6 = 2\pi \times 800\,\text{GHz}\,\mu\text{m}^6$ and $R_c = 7.2 \pm 0.3\,\mu\text{m}$. The right panel is the coherent quantum operations enabled by distance selectivity. **d** Pulse sequence for conditional phase gate implementation using π-pulses (duration 200 ns) and variable delay for interaction. **e** Parity oscillation signal demonstrating coherent interaction between atoms at $R = R_c$, with contrast $\mathcal{C} = 0.85 \pm 0.03$ and phase accumulation rate $2\pi \times 1.2\,\text{MHz}$. **f** Quantum state tomography of generated Bell state $|\Psi^+\rangle = (|gr\rangle + |rg\rangle)/\sqrt{2}$ showing fidelity $\mathcal{F} = 0.88 \pm 0.02$ against ideal state. All error bars represent ± 1 standard deviation from 500 experimental repetitions. Adopted from [28], licensed under CC-BY 4.0 from American Physical Society

For entanglement generation, the protocol involves first preparing the control atom in a superposition state $(|g\rangle + |e\rangle)/\sqrt{2}$ through microwave pulses, followed by conditional evolution under the Rydberg-dressed interaction. The resulting Bell states exhibit concurrence values $C > 0.8$, with the primary limitation coming from residual thermal motion at the experimental temperature. The entanglement verification employs a complete set of correlation measurements, requiring about 10^4 experimental repetitions to achieve sufficient statistical significance for the tomography reconstruction.

The implications of these results extend far beyond the immediate experimental demonstration. The achieved distance-selective interactions open new possibilities for quantum simulation of exotic many-body Hamiltonians, particularly those featuring tunable, long-range interactions such as extended Hubbard models or spin systems with power-law couplings. Moreover, the combination of high-fidelity operations and inherent scalability suggests promising pathways toward fault-tolerant quantum computation with neutral atoms, where the distance selectivity could be exploited for efficient error correction schemes. The techniques developed in this work may also find applications in quantum networking, where controlled interactions between distant atoms could enable the generation of entanglement across multiple quantum nodes.

References

1. D. Kleppner, M.G. Littman, M.L. Zimmerman, Highly excited atoms. Sci. Am. **244**, 54 (1981)
2. M. Marinescu, H.R. Sadeghpour, A. Dalgarno, Dispersion coefficients for alkali-metal dimers. Phys. Rev. A **49**, 982 (1994)
3. M. Saffman, T.G. Walker, Analysis of a quantum logic device based on dipole-dipole interactions of optically trapped Rydberg atoms. Phys. Rev. A **72**, 022347 (2005)
4. M. Saffman, T.G. Walker, K. Mølmer, Quantum information with Rydberg atoms. Rev. Mod. Phys. **82**, 2313 (2010)
5. A.E.B. Nielsen, K. Mølmer, Deterministic multimode photonic device for quantum-information processing. Phys. Rev. A **81**, 043822 (2010)
6. M.D. Lukin et al., Dipole blockade and quantum information processing in mesoscopic atomic ensembles. Phys. Rev. Lett. **87**, 037901 (2001)
7. T.G. Walker, M. Saffman, Consequences of Zeeman degeneracy for the van der Waals blockade between Rydberg atoms. Phys. Rev. A **77**, 032723 (2008)
8. M. Saffman, T.G. Walker, Creating single-atom and single-photon sources from entangled atomic ensembles. Phys. Rev. A **66**, 065403 (2002)
9. T.F. Gallagher, P. Pillet, Dipole-dipole interactions of Rydberg atoms. Adv. At., Mol., Opt. Phys. **56**, 161 (2008)
10. T. Förster, Zwischenmolekulare Energiewanderung und Fluoreszenz. Ann. Phys. **437**, 55 (1948)
11. T.G. Walker, M. Saffman, Zeros of Rydberg-Rydberg Förster interactions. J. Phys. B **38**, S309 (2005)
12. W. Li, P.J. Tanner, T.F. Gallagher, Dipole-dipole excitation and ionization in an ultracold gas of Rydberg atoms. Phys. Rev. Lett. **94**, 173001 (2005)
13. T. Pohl, P.R. Berman, Breaking the dipole blockade: Nearly resonant dipole interactions in few-atom systems. Phys. Rev. Lett. **102**, 013004 (2009)
14. M. Müller et al., Mesoscopic Rydberg gate based on electromagnetically induced transparency. Phys. Rev. Lett. **102**, 170502 (2009)

15. H. Weimer, M. Müller, I. Lesanovsky, P. Zoller, H.P. Büchler, A Rydberg quantum simulator. Nat. Phys. **6**, 382 (2010)
16. K. Singer et al., Suppression of excitation and spectral broadening induced by interactions in a cold gas of Rydberg atoms. Phys. Rev. Lett. **93**, 163001 (2004)
17. K. Singer et al., Spectroscopy of an ultracold Rydberg gas and signatures of Rydberg-Rydberg interactions. J. Phys. B **38**, S321 (2005)
18. A. Reinhard, T.C. Liebisch, B. Knuffman, G. Raithel, Level shifts of rubidium Rydberg states due to binary interactions. Phys. Rev. A **75**, 032712 (2007)
19. A. Gaëtan et al., Observation of collective excitation of two individual atoms in the Rydberg blockade regime. Nat. Phys. **5**, 115 (2009)
20. E. Urban et al., Observation of Rydberg blockade between two atoms. Nat. Phys. **5**, 110 (2009)
21. W.R. Anderson, J.R. Veale, T.F. Gallagher, Resonant dipole-dipole energy transfer in a nearly frozen Rydberg gas. Phys. Rev. Lett. **80**, 249 (1998)
22. T. Vogt et al., Electric-field induced dipole blockade with Rydberg atoms. Phys. Rev. Lett. **99**, 073002 (2007)
23. T.F. Gallagher, P. Pillet, Dipole-dipole interactions of Rydberg atoms. Adv. At., Mol., Opt. Phys. **56**, 161 (2008)
24. T.J. Carroll et al., Simulations of the dipole-dipole interaction between two spatially separated groups of Rydberg atoms. Phys. Rev. A **80**, 052712 (2009)
25. C.S.E. van Ditzhuijzen et al., Spatially resolved observation of dipole-dipole interaction between Rydberg atoms. Phys. Rev. Lett. **100**, 243201 (2008)
26. T.J. Carroll et al., Angular dependence of the dipole-dipole interaction in a nearly one-dimensional sample of Rydberg atoms. Phys. Rev. Lett. **93**, 153001 (2004)
27. P. Bohlouli-Zanjani, J.A. Petrus, J.D.D. Martin, Enhancement of Rydberg atom interactions using ac Stark shifts. Phys. Rev. Lett. **98**, 203005 (2007)
28. S. Hollerith et al., Realizing distance-selective interactions in a Rydberg-dressed atom array. Phys. Rev. Lett. **128**, 113602 (2022)
29. A. Browaeys, T. Lahaye, Many-body physics with individually controlled Rydberg atoms. Nat. Phys. **16**, 132 (2020)

Chapter 3
Cavity-Rydberg Interaction

Abstract Here the interaction between Rydberg atoms and cavity photons is explored. Starting with the JaynesCummings model, the chapter extends to phenomena specific to Rydberg excitations inside cavities, such as self-induced Rabi oscillations, vacuum-field susceptibility, and blackbody-induced decoherence. Photon Fock states, the operation of a Rydberg maser, and cavity-mediated ultrafast energy exchange are analyzed. Special focus is given to experimental schemes coupling cold Rydberg atoms with cavities, including two-photon excitation and challenges in free-space excitation, emphasizing the potential for strong light-matter coupling at the few-photon level.

3.1 Introduction

The cavity QED possesses a central role to implement and control light-matter interactions. It provides a highly controlled platform where quantized light fields and quantized atomic states can interact coherently over extended timescales. Strongly driven optical cavity with atoms (or an atom) is the best way to engineer strong-coupling regime, as shown in Fig. 1.1, where the atoms inside a strongly driven optical cavity are illustrated. In the strong-coupling regime, the coherent exchange of excitations between atoms and cavity field dominates over dissipation, enabling phenomena such as vacuum Rabi oscillations, photon blockade, and enhanced nonlinear optical processes. Due to the high-Q (high quality factor) nature of the cavity, intra-cavity photons perform many round trips before decaying out of the cavity which not only cause the coherent manipulation of atoms but also facilitate to have strong atom-field coupling with long life-time. This prolonged photon storage time enhances the probability of atom-photon interaction and is critical for applications in quantum memories and repeaters. These features also help to keep intra-cavity atoms at low temperature and provide precise control over the position of atomic mode. In turn, this allows for high-fidelity addressing of atomic states and control over the atom-cavity coupling strength, especially in setups where the position-dependent cavity mode structure is exploited for deterministic interactions.

K. A. Yasir and W.-M. Liu, *Rydberg Atoms in Cavity*, Springer Series in Solid-State Sciences 213, https://doi.org/10.1007/978-981-95-4571-1_3

The inclusion of Rydberg atoms inside the cavity generates long-range dipole-dipole interactions which lead to the strong blockade excitations. These interactions extend far beyond the typical range of van der Waals forces and allow one excited Rydberg atom to influence the excitation probability of atoms several micrometers away. In such cases, only one Rydberg excitation is allowed within a given blockade radius, leading to collective excitation states that significantly differ from single-atom dynamics. The intra-cavity optical mode also assists in achieving and sustaining long life-time Rydberg states. It not only helps to excite the atoms into high principal quantum number states with minimized spontaneous decay but also stabilizes their interaction through reduced blackbody-induced ionization and collisional effects. As, in blockade regime, an ensemble of atoms can be modeled in the form of two-level system, therefore, such cavity induced long-range interactions enhance the significance and experimental feasibility of cavity-Rydberg systems. This effective two-level approximation simplifies the quantum dynamics, making it more amenable to analytical modeling and scalable quantum control schemes. Moreover, the collective enhancement of atom-light coupling scales with the square root of the number of participating atoms, which further strengthens the nonlinear optical response of the system.

Thus, cavity QED with Rydberg atoms is very important in order to text and implement various phenomena of quantum physics, especially related to quantum information process. It offers a versatile platform for realizing deterministic quantum gates, single-photon transistors, quantum phase transitions, and entangled state preparation. These systems also serve as testbeds for exploring strongly correlated photonic matter, where photons behave like interacting particles due to the effective nonlinearity introduced by the Rydberg blockade. The tunability of both cavity parameters and atomic interactions allows researchers to simulate many-body quantum systems, investigate topological photonic states, and probe quantum critical phenomena under controlled laboratory conditions.

We have discussed the basic properties as well as interactions which are necessary to understand and implement Rydberg atoms in cavities in last section. In this chapter, we will review the fundamental modeling of Rydberg atoms inside cavity and cavity mediated controlled excitations of Rydberg atoms. The upcoming sections will focus on the Hamiltonian formulation, quantized interaction dynamics, and various configurations of cavity-Rydberg coupling schemes–including single-atom, few-atom, and collective excitation regimes–along with their respective implications for quantum simulation and quantum technological applications.

3.2 The Jaynes-Cumming Model

The Jaynes-Cumming model (JCM) was proposed by Edwin Jaynes and Fred Cummings in 1963 in which they developed an interface between quantum mechanical and semi-classical definitions of electromagnetic radiation in order to demonstrate spontaneous emission of radiation [1]. The semi-classical picture defines matter-wave interaction by only considering quantized form of matter while electromagnetic radiations are defined by a definite function instead of quantized operator. This concept has been very useful to describe many features of modern optics but regarding atom-field interactions and their application, the semi-classical theory appears to be useless. The inability of the semi-classical theory to account for discrete energy exchange and entanglement between light and matter fundamentally limits its explanatory power in regimes involving strong coupling or quantum coherence. However, the JCM also uses quantized electromagnetic radiation to study the matter-wave interaction by evolving two-level systems, in comparison with semi-classical picture. In this model, both the atom and the field are treated quantum mechanically, allowing for the description of phenomena such as Rabi oscillations, collapse and revival of atomic populations, photon antibunching, and quantized energy exchange.

The JCM has explained many quantum mechanical phenomena which were limited by the semi-classical picture. Such as, the revival of the quantum mechanical wave-packet after collapse, which appears to be the direct consequence of quantized electromagnetic field [2]. This phenomenon arises due to the discrete nature of the photon number states and their contribution to the time evolution of the coupled atom-field system, revealing the intricate coherence between atomic and photonic subsystems. The prediction and experimental confirmation of these effects have laid the foundation for cavity quantum electrodynamics (QED) and opened pathways for the development of quantum information processing platforms. Regarding atom-cavity system, JCM has played an important role to visualize strong atom-field coupling and the behavior of optical mode inside the cavity. It has also provided critical insight into the nature of dressed states and normal-mode splitting under resonant and near-resonant conditions. These features are essential for designing systems that require deterministic control over single photons and atoms, such as quantum gates and photon-mediated entanglement.

The JCM is so far a best way to develop and evaluate hybrid systems especially for two-level atomic in cavities. It serves as a fundamental tool to describe coherent dynamics, quantum state engineering, and dissipative effects in systems ranging from cavity QED and circuit QED to trapped ion platforms. The model can be extended to include multiple atoms (Tavis-Cummings model), time-dependent parameters (driven JCM), or coupling to environmental modes (dissipative JCM), which enhances its relevance in real-world quantum technological settings. As Rydberg atoms can be modeled as two-level system by taking care of strong interaction regime, therefore, JCM is the perfect formalism to describe their evolution dynamics inside the cavity. The strong and tunable dipole-dipole interactions between Rydberg states allow for an effective two-level description under blockade conditions, aligning well with the assumptions of the JCM. Furthermore, in cavity-Rydberg platforms, where single or few atoms interact with a quantized optical mode, the JCM not only captures the essential quantum dynamics but also enables the study of nonlinear optical effects, photon blockade, and quantum phase transitions. Thus, the Jaynes-Cummings framework provides a powerful and intuitive basis for modeling and understanding the rich physics of Rydberg atoms in cavity QED environments.

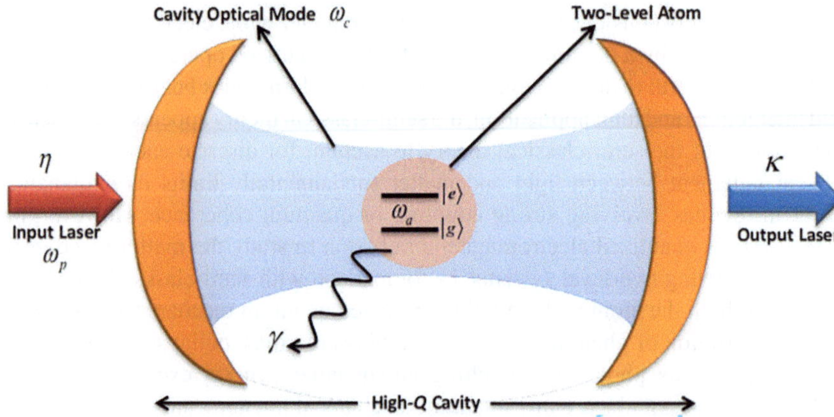

Fig. 3.1 The schematic diagram for the cavity-Rydberg system. The two-level Rydberg atom interacts with intra-cavity field oscillating at frequency ω_c and driven by single mode optical field ω_p. Here ω_a is the transitional Rabi frequency between ground state $|g\rangle$ and excited state $|e\rangle$ while κ is the cavity photon decay. η is the coupling of intra-cavity field with external driving field and γ is the decay of intra-cavity atomic excited state

Here we develop a JCM for a Rydberg two-level system trapped inside a high finesse cavity, or a Fabry-Pérot cavity, driven by the single mode external laser with frequency ω_p, as shown in Fig. 3.1. Due to the high finesse and quality factor, the photon inside the cavity performs a large number of round trips before leaving the cavity. This will create a coherent optical mode inside the cavity oscillating at frequency ω_c, which will interact and excite two-level atomic system with transition frequency ω_a. The energy of a photon inside a cavity mode is defined by $\hbar\omega_c$, which is equivalent to the spacing between intra-cavity energy levels. The intra-cavity optical mode with n number of photons form energy eigenstate $|n\rangle$ having a counterpart $|0\rangle$, which represents vacuum state of no photon state. The two-level configuration of Rydberg atoms inside the cavity forms a Hilbert space composed of two energy eigenstates, a ground state $|g\rangle$ and an excited state $|e\rangle$. Because of having two-dimensional Hilbert space, these states of Rydberg atoms can be modeled as vectors, defined as,

$$\alpha|e\rangle + \beta|g\rangle = \begin{pmatrix} \alpha \\ \beta \end{pmatrix}, \tag{3.1}$$

where α and β are the probability amplitudes for finding atom in excited state and in ground state, respectively, defined by the maximum probability relation $|\alpha|^2 + |\beta|^2 = 1$. The frequencies of cavity mode as well as atomic transition are detune by the external pump field frequency as $\Delta_c = \omega_p - \omega_c$ and $\Delta_a = \omega_p - \omega_a$.

The total Hamiltonian of the atom-cavity system consists of three parts,

$$\hat{H}_{JM} = \hat{H}_{field} + \hat{H}_{atom} + \hat{H}_{int} + \hat{H}_{in}, \tag{3.2}$$

where $\hat{H}_{field}$ corresponds to the cavity mode, $\hat{H}_{atom}$ represents Rydberg atom, $\hat{H}_{int}$ accumulates atom-atom and atom-field interactions and $\hat{H}_{in}$ provides relation of intra-cavity field with external pump field. The Hamiltonian for cavity field can be written as,

$$\hat{H}_{field} = \hbar\Delta_c \hat{a}^\dagger \hat{a}, \tag{3.3}$$

where $\hat{a}^\dagger$ and $\hat{a}$ are the photon creation and annihilation operators for the cavity mode, respectively, defined by the quantized electric field $\mathbf{E}(r, t)$ and magnetic field $\mathbf{B}(r, t)$ vectors. Here $\hbar$ is the quantization constant or Planck constant. The commutation relation for photonic creation and annihilation operators can be defined as $[\hat{a}^\dagger \hat{a}] = i$. The Hamiltonian for Rydberg two-level atom inside cavity can be illustrated as,

$$\hat{H}_{atom} = \frac{\hbar\Delta_a(r)}{2}\hat{\sigma}_z, \tag{3.4}$$

where $\Delta_a(r) = \Delta_a + \Delta_s(r)$ is the position dependent detuning of the Rydberg atoms and r describes the position of atoms inside cavity. Here $\Delta_s(r)$ accounts for the ac-Stark shift which is induced by the far-detuned intra-cavity optical trap. The operator $\hat{\sigma_z} = |e\rangle\langle e| + |g\rangle\langle g|$ is the atomic inversion operator describing the state of two-level atom. To illustrate atom-field interaction, we use interaction picture instead of Schrödinger picture. In this regard, the Hamiltonian for atom-field interaction $\hat{H}_{int}$ can be written as,

$$\begin{aligned}\hat{H}_{int} = {} & \frac{\hbar U_0}{2}(\hat{a}\hat{\sigma}_- e^{-i(\omega_c+\omega_a)t} + \hat{a}^\dagger\hat{\sigma}_+ e^{i(\omega_c+\omega_a)t} + \hat{a}\hat{\sigma}_+ e^{i(-\omega_c+\omega_a)t} \\ & + \hat{a}^\dagger\hat{\sigma}_- e^{-i(-\omega_c+\omega_a)t}),\end{aligned} \tag{3.5}$$

where $\hat{\sigma}_+ = |e\rangle\langle g|$ and $\hat{\sigma}_- = |g\rangle\langle e|$ are the raising and lowering ladder operators for atomic mode, like operator $\hat{a}$ and $\hat{a}^\dagger$, describing the transition of atom from excited level to ground level and ground level to excited level, respectively. The two-level Rydberg atom trapped in cavity pretends like 1/2-spin system whose transition can be described by three-dimensional Bloch vector in Block sphere. Here it should be noted that such two-level systems actually do not have spin but they behave like two 1/2-spin systems due to their two-dimensional Hilbert space. Further, $U_0 = g_0/\Delta_a$ is the depth of optical potential for atom. It can also be defined as the coupling of atoms with intra-cavity field, which means by increasing the depth of potential atoms will be more coupled to cavity mode. Here $g_0 = d\sqrt{\hbar\omega_c/2\epsilon_0\mathcal{V}}$ is the single-photon Rabi frequency, where parameter d represents atomic dipole moment along the intra-cavity field polarization and $\mathcal{V}$ is the cavity mode volume. The exponential functions in Eq. (3.5) represent cavity mode function defining the configuration of cavity depending upon time and spatial distribution of atomic mode. The cavity mode function can vary depending upon the intra-cavity optical field, like for standing wave configuration and running wave configuration, cavity mode function will be different. Eq. (3.5) contains both quickly moving $\omega_c + \omega_a$ and slowly moving

$\omega_c - \omega_a$ components of intra-cavity potential. In order to obtain solvable model, the counter-rotating components for the quickly oscillating term can be ignored $|\omega_c - \omega_a| << \omega_c + \omega_a$ by applying condition $\omega_p >> g_0, \Delta_c, \Delta_a$. This approximation is known as the rotating-wave approximation and has been frequently used in such cavity systems. By using rotating-wave approximation, the Hamiltonian for atom-field interaction in Schrödinger picture will appear as,

$$\hat{H}_{int} = \frac{\hbar U_0}{2}(\hat{a}\hat{\sigma}_+ + \hat{a}^\dagger\hat{\sigma}_-). \tag{3.6}$$

Finally, the Hamiltonian describing relation of intra-cavity field and external pump field can be written as, $\hat{H}_{in} = i\hbar\eta(\hat{a}^\dagger - \hat{a})$, where $|\eta| = \sqrt{P\kappa/\hbar\omega_p}$ is the coupling of intra-cavity field with external pump field. The parameter P is the power of external pump field and κ is the cavity photon decay rate. The Hamiltonian $\hat{H}_in$ is explicitly time dependent because we consider the rotating-wave approximation over the frequency ω_p of external driving field.

In order to calculate eigenstates and eigenenergies corresponding to these states, we divide Jaynes-Cumming Hamiltonian (or JC Hamiltonian) $\hat{H}$ in to two parts $\hat{H}_{JC} = \hat{H}_1 + \hat{H}_2$. By excluding external pump field part for simplicity and imposing rotating-wave and electric-dipole approximation, $\hat{H}_1$ and $\hat{H}_2$ can be defined as,

$$\hat{H}_1 = \hbar\omega_c(\hat{a}^\dagger\hat{a} + \frac{\hat{\sigma}_z}{2}), \tag{3.7}$$

$$\hat{H}_2 = \hbar\Delta_a\frac{\hat{\sigma}_z}{2} + \frac{\hbar U_0}{2}(\hat{a}\hat{\sigma}_+ + \hat{a}^\dagger\hat{\sigma}_-). \tag{3.8}$$

The eigenstates describe the state of the two-level atom both in excited or ground levels. For Hamiltonian $\hat{H}_1$, the eigenstates can be easily obtained with the help of tensor product rule and can be denoted as $|\hat{\psi}_1\rangle := |n, e\rangle$ and $|\hat{\psi}_2\rangle := |n, g\rangle$, where n is the total number of photons inside the cavity. These states lead to diagonalization of the JC Hamiltonian $\hat{H}_{JC}$ over the subspace $\{|\hat{\psi}_1\rangle, |\hat{\psi}_2\rangle\}$ as, $\hat{H}_{ij}^{(n)} = \langle\hat{\psi}_{in}|\hat{H}_{JC}|\hat{\psi}_{jn}\rangle$, where i and j correspond to the eigenstates at n and $\hat{H}_{ij}^{(n)}$ describes Hamiltonian of that particular state. The matrix elements corresponding to $\hat{H}_{JC}$ over the subspace $\{|\hat{\psi}_1\rangle, |\hat{\psi}_2\rangle\}$ are read as,

$$\hat{H}^{(n)} = \hbar\begin{pmatrix} n\omega_c + \frac{\omega_a}{2} & \frac{U_0}{2}\sqrt{n+1} \\ \frac{U_0}{2}\sqrt{n+1} & (n+1)\omega_c - \frac{\omega_a}{2} \end{pmatrix}. \tag{3.9}$$

The energy eigenvalues, as a function of photon number n, can be obtained by simply calculating eigenvalue of above matrix, given as,

$$E_\pm(n) = \hbar\omega_c(n + \frac{1}{2}) \pm \frac{1}{2}\hbar\Omega_n(\Delta_a), \tag{3.10}$$

where $\Omega_n(\Delta_a) = \sqrt{\Delta_a^2 + U_0^2(n+1)}$ is the effective Rabi frequency, as a function of atomic transition detuning Δ_a, corresponding to n. The eigenstates for these energy eigenvalues can be written as,

$$|n, -\rangle = cos(\frac{\alpha_n}{2})|\hat{\psi}_{1n}\rangle + sin(\frac{\alpha_n}{2})|\hat{\psi}_{2n}\rangle, \tag{3.11}$$

$$|n, -\rangle = -sin(\frac{\alpha_n}{2})|\hat{\psi}_{1n}\rangle + cos(\frac{\alpha_n}{2})|\hat{\psi}_{2n}\rangle, \tag{3.12}$$

where angle $\alpha_n = tan^{-1}(U_0\sqrt{n+1}/\Delta_a)$ defines the phase of the Rydberg atoms.

Further, it is important to discuss the time evolution of the atom-cavity system with the help of JCM. For this purpose, conventionally, two approaches are used regarding system configuration. One is Schrödinger picture which describes time evolution by following Schrödinger equation and the other one is Heisenberg picture which is based on unitary operator method. They both have served very well in order to interpret temporal dynamics of the systems. In Schrödinger picture, time evolution of the system can be obtained by expanding general state of the atoms into eigenstates depending upon the value of n. For this, first we describe the initial state of the cavity mode by using quantum superposition principle, $|\hat{\psi}_F(0)\rangle = \sum_n C_n|n\rangle$, where $|n\rangle$ is the number state corresponding to the intra-cavity photon number and C_n is probability amplitude for that particular number state. It can also be called as state vector defining the element of complex Hilbert space and their sum of absolute square must obey $\sum C_n^* C_n = 1$. By assuming trapped atom in excited state, the initial state of whole atom-cavity system can be illustrated as,

$$|\psi(0)\rangle = \sum_n C_n[cos(\frac{\alpha_n}{2})|n, +\rangle - sin(\frac{\alpha_n}{2})|n, -\rangle]. \tag{3.13}$$

By using this initial state in the solution of time dependent Schrödinger equation with JC Hamiltonian, we can write the final time dependent form of the system state,

$$\begin{aligned} |\psi(t)\rangle &= e^{-i\hat{H}_{JC}t/\hbar}|\psi(0)\rangle \\ &= \sum_n C_n[cos(\frac{\alpha_n}{2})|n, +\rangle e^{-iE_+(n)t/\hbar} \\ &\quad -sin(\frac{\alpha_n}{2})|n, -\rangle e^{-iE_-(n)t/\hbar}]. \end{aligned} \tag{3.14}$$

As the sinusoidal wave functions are dependent on effective Rabi frequency which is the function of photon number, therefore, for each photon number state, there will be a wave function. The destructive sum of these wave functions over the time can approach to zero depending upon the nonlinearities but later in time, that wave functions start reappearing because of quantum interference among these wave functions at difference eigenstates. This phenomenon of quantum interference is in the foundation of many quantum mechanical phenomena.

However, in Heisenberg picture, the time evolution is calculated by defining unitary operator for JC Hamiltonian as,

$$\hat{U}(t) = e^{-i\hat{H}_{JC}t/\hbar}$$
$$= \begin{pmatrix} e^{-i\omega_c(\hat{a}^\dagger \hat{a}+\frac{1}{2})t}(cost\sqrt{\hat{\phi}+U_0^2}-i\frac{\Delta_a}{a}\mathbf{A}) & -iU_0 e^{-i\omega_c(\hat{a}^\dagger \hat{a}+\frac{1}{2})t}\hat{a}\mathbf{A} \\ -iU_0 e^{-i\omega_c(\hat{a}^\dagger \hat{a}-\frac{1}{2})t}\hat{a}^\dagger\mathbf{B} & e^{-i\omega_c(\hat{a}^\dagger \hat{a}-\frac{1}{2})t}(cost\sqrt{\hat{\phi}}+i\frac{\Delta_a}{a}\mathbf{B}) \end{pmatrix}, \quad (3.15)$$

where $\mathbf{A} = sint\sqrt{\hat{\phi}+U_0^2}/\sqrt{\hat{\phi}+U_0^2}$, $\mathbf{B} = sint\sqrt{\hat{\phi}}/\sqrt{\hat{\phi}}$ and operator $\hat{\phi} = U_0^2\hat{a}^\dagger + \Delta_a^2/4$. Now by using initial density matrix, like initial photon state, $\hat{\rho}(0)$ and the initial expectation value of the observable $\langle\hat{\Theta}\rangle$, we determine the temporal behavior of the atomic mode inside the cavity, given as,

$$\hat{\rho}(t) = \hat{U}^\dagger(t)\hat{\rho}\hat{U}(t), \quad (3.16)$$
$$\langle\hat{\Theta}\rangle_t = Tr[\hat{\rho}(t)\hat{\Theta}]. \quad (3.17)$$

The JCM has been extensively used to understand and implement cavity-Rydberg systems in almost all theoretical as well as experimental investigation. In following discussion JCM will play a fundamental rule to describe cavity-Rydberg systems.

3.3 Rydberg Excitations in Cavity

The inclusion of Rydberg atoms inside a high-Q cavity triggers a strong coupling between atomic degree of freedom and cavity optical degree of freedom, which will lead to the energy transformation between atoms and cavity field through absorption and emission mechanism. This strong coupling regime marks a significant departure from weak coupling, as it allows coherent exchange of quanta between atoms and the field, rather than mere irreversible decay. When excited state atoms $|e\rangle$ interact with an empty cavity, having a vacuum state or zero photon state $|0\rangle$, then they may spontaneously decay to the ground state $|g\rangle$ after emitting a finite number of photons to the cavity, which will build an intra-cavity photon number state $|n\rangle$. In this framework, the atomic excitation energy is transferred into quantized radiation within the cavity, leading to discrete steps in the photon number distribution that can be experimentally detected. Such exchange of energy between atoms and intra-cavity field occurs at rate of $g_0 = 2d\mathcal{E}_0/\hbar$, which is known as Rabi oscillation, where parameter d is the electric dipole moment and $\mathcal{E}_0 = \sqrt{\hbar\omega_a/2\epsilon_0\mathcal{V}}$ is the electric field intensity for single photon inside cavity with vacuum state volume $\mathcal{V}$. These oscillations signify a coherent back-and-forth energy transfer between atom and cavity mode, and they serve as the hallmark of quantum coherence in cavity QED systems.

Such kind self-induced Rabi oscillations are only possible when their lifetimes are in the range greater than the decay rates of intra-cavity photon and atom $1/T_c$ and $1/T_a$, respectively. In other words, the time scales governing photon leakage from the cavity and atomic spontaneous emission must be much longer than the Rabi oscillation period to preserve coherence. The intra-cavity photon decay time is directly proportional to the value of the high-Q (high quality) factor $T_c = Q/\omega_c$, which helps us to develop condition for Rabi oscillation $T_c g_0 > 1$. This inequality implies that the system enters a strong coupling regime only if the coherent coupling rate g_0 dominates over the relevant decay rates. When satisfied, this condition enables the observation of vacuum Rabi splitting in the transmission spectrum and permits the implementation of photon-mediated quantum gates. The experimental demonstration of such systems that obey this condition with high-Q factor have given initially tough time to the researcher. Challenges included suppressing thermal fluctuations, isolating single atoms or small ensembles, and maintaining cavity stability over long timescales. Furthermore, Rydberg atoms, due to their large size and sensitivity, required specially designed trapping and shielding techniques to minimize decoherence.

However, it was crucially important to achieve this condition in order study various phenomena of modern physics. Realizing strong coupling between Rydberg atoms and cavity fields has allowed for deeper exploration of non-classical light generation, quantum entanglement, photon blockade, and quantum phase transitions in engineered optical environments. These systems are now being used to simulate interacting many-body quantum systems, develop quantum logic operations, and even investigate the interplay of long-range interactions and cavity-mediated correlations. In particular, combining the large electric dipole moments of Rydberg atoms with the enhanced field confinement of high-Q cavities enables nonlinear effects at the single-photon level, bridging the gap between quantum optics and quantum information science. As a result, cavity-Rydberg systems stand as a promising platform for implementing scalable quantum technologies with high fidelity and robustness.

3.3.1 Self-Induced Rabi Oscillations

To observe self-induced Rabi oscillation with atom-field interaction in cavity, it is necessary to fulfill the condition $T_c g_0 > 1$, which is only possible in high-Q regime. This condition signifies that the coherent interaction time between atom and field must exceed the characteristic decay time of the cavity mode to allow multiple coherent energy exchanges. If this condition is not satisfied then the emitted photon from atom will decay out from the cavity mirrors, because of having low reflection index, instead of being reabsorbed by the atoms again. In such a scenario, the atom-field system fails to reach the strong coupling regime, and the dynamics become dominated by dissipation and incoherent decay. At this point, intra-cavity field will behave as an irreversible photonic sink with decay rate $\Gamma_c = g_0^2 T_c$, which is much larger than the spontaneous emission itself. This enhanced decay mechanism leads to the suppression of coherent Rabi oscillations and causes the system to mimic a lossy environment, where the quantum coherence is quickly lost. The enhancement in spontaneous emission by self-induced Rabi oscillation, in high-Q cavities, was first theoretically predicted by the group of E. M. Purcell in 1946 [4] and was later experimentally confirmed by the research group of S. Haroche in 1983 [5].

In their experiment, they demonstrated enhanced spontaneous emission of radiation in cavity-Rydberg system. They used Rydberg atoms in a high-Q superconducting Fabry-Pérot cavity with electromagnetic field oscillating at 340 GHz. These Rydberg atoms were excited to high principal quantum numbers, thereby possessing exaggerated electric dipole moments, which significantly increased their coupling strength to the cavity mode. In Fabry-Pérot cavities, it is possible to generate volume of cavity mode greater than the transition wavelength. This condition is crucial because the spatial extent of the cavity mode influences the density of optical states and hence the atom-field coupling strength. Further, it is easy to increase the Q factor by using highly refined mirrors. High reflectivity superconducting mirrors can reduce photon leakage from the cavity, thus enhancing the effective photon lifetime and allowing multiple absorption–emission cycles between atoms and the cavity mode.

Thus, it was possible to engineer strong coupling regime in cavity-Rydberg system because of high finesse cavity and strong electric dipole moment of Rydberg atoms, which increase the life-time of Rydberg atoms inside the cavity. The combination of high-Q factor, low thermal noise, and ultra-cold atoms allowed for a quantum electrodynamical platform where the coherent dynamics could be observed and precisely controlled. Moreover, the Purcell effect in such systems not only enhances spontaneous emission but also modifies the spectral and temporal characteristics of emitted photons, enabling the generation of indistinguishable single photons and nonclassical light states. These achievements laid the groundwork for subsequent developments in cavity QED, including photon entanglement, atom-photon gate operations, and the construction of prototype quantum networks based on cavity-coupled Rydberg systems. Today, the principles established by these pioneering experiments are instrumental in developing scalable architectures for quantum computation and communication using strongly interacting light-matter systems.

The group of S. Haroche demonstrated the cavity mediated enhancement in spontaneous emission by measuring ionization of atoms leaving the cavity as a function of time, as shown in Fig. 3.2a. The dotted and solid curves correspond to the non-resonant and resonant configuration of the cavity, respectively. The notions a, b and c represent the average number of ionized electrons 0.35, 0.2 and 0.13 corresponding to the average of Rydberg atoms 3.5, 2 and 1.3 detected from the $23s$ state of Na atoms, respectively. One can distinguish the difference between ionization spectrum of off-resonant and resonant cavity configurations. The atoms inside cavity get more ionized when the cavity mode is resonant because of interacting with more photons. Further, due to high-Q factor, number of passing atoms have also been increased, which can be observed with increase in ionization. The curvature of solid curves corresponds to the superposition principle, as a function of time, which can be defined as $\Pi_{sp}(t) = (1 - \mathbf{x})\Pi_s(t) + \mathbf{x}\Pi_p(t)$, where $\Pi_s(t)$ and $\Pi_p(t)$ are the probability for finding atomic states $23s$ and $22p$, respectively, while $\mathbf{x}$ is the number of atoms inside the cavity. In their case, the enhancement parameter was $\mathbf{x} = 0.15 \pm 0.05$, depending upon off-resonant and resonant cavity configurations, which yield in quality factor

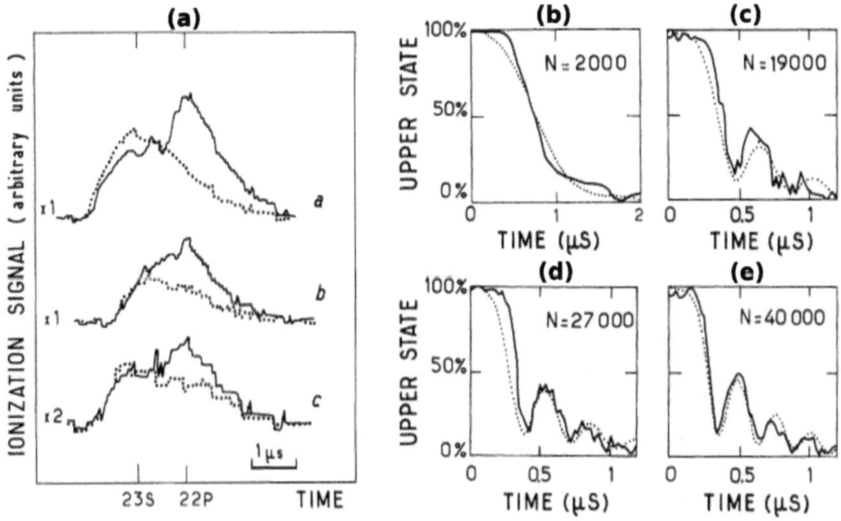

Fig. 3.2 **a** Cavity mediated enhanced spontaneous emission obtained from ionization signal as a function of time. The solid lines correspond to the resonant cavity configuration while dotted lines accommodate off-resonant response of the system. The traces a, b and c represent average intra-cavity atomic number 3.5, 2 and 1.3, respectively. a and b lines are for $23s \rightarrow 22p_{3/2}$ and the line c is for $23s \rightarrow 22p_{1/2}$ pair states of Rydberg atoms. Obtained from [5], ©(2025) with permission from American Physical Society. **b–e** The intra-cavity upper level population spectrum as a function of running time for intra-cavity atomic number $N = 2000$, 19000, 27000 and 40000, respectively. The solid lines are for experimental results while dotted lines are for the theoretical calculations. Obtained from [6], ©(2025) with permission from American Physical Society

$Q \approx 7.5 \times 10^5$. The value of Q factor was appear to be limited because of cavity mediated Stark shift, induced by the small components of incident electromagnetic field. Further, if we increase the time of interaction between intra-cavity optical field and atoms (more than one) then the intra-cavity photons will not only get coupled with all atoms but will also generate mutual coupling between atoms causing superradiant effects in atoms. This cavity-assisted superradiant dynamics in Rydberg atoms have been extensively studied in such cavity-atom systems, as discussed in following sections.

In another experiment, performed by the group of S. Haroche, the self-induced Rabi oscillations for Rydberg atoms in an optical cavity and their dependence on Q factor of the cavity were studied [6]. The main purpose of this experiment was to demonstrate that if N number of Rydberg atoms emit radiation after interacting with cavity field whose strength is larger than the $\sqrt{N}\mathcal{E}_0$ then the effective Rabi transition will be the order of $g_0\sqrt{N}$. Here $\mathcal{E}_0$ is the electric field intensity for one photon. It will modify the condition for cavity Q factor in the form of $g_0\sqrt{N}T_c > 1$, which is easier to achieve even with low photon life-time T_c. They also provide an important comparison between theoretical and experimental results, as can be seen in Fig. 3.2b, where the population of excited Rydberg atoms inside the cavity versus time has been shown. The solid and dotted curves correspond to the experimental and theoretical results, respectively. It can be observed that with relatively small number of atoms inside the cavity, the condition for enhanced spontaneous emission has not been fulfilled, therefore, atoms could not reabsorb emitted photons which lead to a smooth decay in the atomic population. Figure 3.2b shows these effects, where atomic excited state population is $N = 2000$. This case is suitable for cavity mediated over-damping which eventually causes superradiant regime. However, when we increase the number of excited state atoms, which can be calculated with formula $N_e/(N_e + N_g)$, the intra-cavity atoms are able to reabsorb photons emitted by them after interacting with cavity field. By increasing the number of excited state atoms, the net amount of radiated photons inside the cavity will be increased, which consequently causes the emergence of self-induced Rabi oscillations. One can observe these effects in results shown in Fig. 3.2b, where the excited atomic population is increased to $N = 19000$, $N = 27000$ and $N = 40000$, respectively. Here, Rydberg atoms play an important role due to their extraordinary features, including large distance between ground level and excited level, which increases the strength and lifetime of effective Rabi oscillations.

Further, the enhanced spontaneous emission through self-confinement of Rydberg atoms inside cavity has been discussed in strong-coupling regime [7] and in time dependence manipulation of Rabi oscillation to elaborate collapse-revival phenomenon [8]. The entanglement and decoherence of Rydberg ensemble inside cavity with enhanced spontaneous emission have been investigated [9]. These investigations are particularly important in the context of quantum information processing and quantum memories, where control over decoherence is essential. Recently, a connection between Grassmann phase space theory and JCM for cavity-Rydberg system was developed by manipulating intra-cavity Rabi oscillations [10]. They modeled Grossmann variable as an intra-cavity optical mode to interpret JCM for cavity-Rydberg system in order to understand the phase space dynamics of the system, which can play an essential role to understand the classical-quantum interface. Moreover, these advancements also serve as a fundamental stepping stone toward the realization of quantum simulators and hybrid cavity-QED systems, where Rydberg atoms can act as a bridge between optical and microwave quantum platforms due to their tunable coupling strengths and long coherence times. This ability to tailor atom-field interactions through cavity QED paves the way for engineering quantum states of light and matter at will, ultimately advancing the frontiers of quantum optics, quantum thermodynamics, and fundamental tests of quantum theory. These investigations build a foundations to test and implement millimeter and micrometer cavities with Rydberg ensembles and to demonstrate controlled atom-field interactions.

3.3.2 Vacuum-Field Susceptibility

The behavior of cavity-Rydberg systems are not only influence by the intra-cavity photon number states, or cavity field, but also depend on the intra-cavity vacuum mode, which shows significant effects on the absorption and emission spectrum of Rydberg atoms. Therefore, it is important to discuss fundamental effects of matter-wave interactions in presence of cavity vacuum state, which causes vacuum-field Rabi oscillation. Further, the susceptibility of strongly interacting Rydberg ensembles inside the cavity will be modified after interacting with intra-cavity optical mode, even if the optical mode of the cavity is in vacuum state, or no photon state. Here we discuss such modification in susceptibilities due to intra-cavity vacuum states, as theoretically demonstrated in Ref. [11], where authors discussed the effects of intra-cavity vacuum state on Rabi oscillations by measuring absorption spectrum. They consider a Rydberg ensemble of N two-level atoms inside the cavity with transitional frequency ω_a, as shown in Fig. 3.1. By characterizing two-levels of atoms as a 1/2-spin operators, $S_i^{\pm}$ and S_i^z, and assuming their effective strength as, $\sum_i S_i = S$, the effective JC Hamiltonian for system will read as,

$$\hat{H} = \hbar\omega_a S^z + \hbar\omega_c \hat{a}^{\dagger}\hat{a} + (\hbar g S^{+}\hat{a} + H.c.), \tag{3.18}$$

where g is the strength of coupling between atoms and intra-cavity optical mode defined by the intra-cavity potential depth and ω_c is the frequency of cavity mode. As the atoms are consider in initially in ground state and cavity field in vacuum state $|0\rangle$, therefore, corresponding coherent state of the system can be written as $|\psi_0\rangle = |N/2, -N/2; 0\rangle$ with energy $E_0 = -\hbar N\omega_0/2$. It illustrates the excitations induced by the cavity vacuum state. The intra-cavity photon decay rate κ must not be large scale otherwise vacuum Rabi oscillations will be suppressed. In this regard, they developed a condition for cavity decay rate by using same condition for enhanced spontaneous emission condition $g\sqrt{N} >> \kappa$, as discussed in previous section. By applying JCM, the density matrix for the cavity-Rydberg system can be formalized as,

$$\frac{\partial \hat{\rho}}{\partial t} = -\frac{i}{\hbar}[\hat{H}, \rho] - \kappa(\hat{a}^{\dagger}\hat{a}\rho - 2\hat{a}\rho\hat{a}^{\dagger} + \rho\hat{a}^{\dagger}\hat{a}). \tag{3.19}$$

By utilizing eigenstates and corresponding energy eigenvalues and imposing secular approximation [12], which satisfies Pauli-type master equation, the off-diagonal elements of density matrix will be read as,

$$\frac{\partial \hat{\rho}_{ij}}{\partial t} = -\frac{i}{\hbar}(E_i - E_j)_{\rho_{ij}} - \Gamma_{ij\rho_{ij}}, \tag{3.20}$$

where parameter Γ depends upon the structural incoherent terms which can also be interpreted as a system damping factor.

By using Eq. (3.20), the susceptibility $\mathcal{X}_{\alpha\beta}$ as a function of eigenfunctions and energy eigenvalues, in presence of external electromagnetic field ω_p, can be expressed as,

$$\mathcal{X}_{\alpha\beta}(\omega_p) = d_\alpha d_\beta N \sum_i \frac{|N/2, N/2+1; 0|\psi_i\rangle|^2}{(E_j - E_0 - \hbar\omega_p - i\hbar\gamma_{i,0})} + terms\ with - \omega_p, \quad (3.21)$$

where d_α and d_β are the αth and βth components of the dipole matrix (or electric dipole moment matrix). As the eigenfunction of the Hamiltonian is characterize by the eigenvalues corresponding to S^2 and $S^z + \hat{a}^\dagger\hat{a} = C$, therefore, these particular eigenstates and corresponding energy eigenvalues can be written as,

$$|\psi_-\rangle = cos\theta|n/2, -N/2+1; 0\rangle + sin\theta|n/2, -N/2; 1\rangle, \quad (3.22)$$

$$|\psi_-\rangle = -sin\theta|n/2, -N/2+1; 0\rangle + cos\theta|n/2, -N/2; 1\rangle \quad (3.23)$$

$$E_\pm = \hbar\omega_a(-\frac{N}{2}+1) - \hbar\frac{\Delta_a}{2} \pm \frac{1}{2}\sqrt{\Delta_a^2 + 4Ng^2}, \quad (3.24)$$

where $\Delta_a = \omega_a - \omega_c$ is the effective transitional detuning and $\theta = tan^{-1}(-\Delta_a + \sqrt{\Delta_a^2 + 4Ng^2}/2g\sqrt{N})$ is the angle of interaction between atoms and intra-cavity optical mode. By using these definitions in Eq. (3.21), the resultant susceptibility can be rewritten as, $\mathcal{X}_{\alpha\beta} = (\pi N d_\alpha d_\beta/\hbar)\mathcal{X}(\omega_p)$, whose imaginary part will read as,

$$Im\mathcal{X}(\omega_p) = cos^2\theta \frac{\frac{\Gamma_-}{\pi}}{\Gamma_-^2 + [\omega_p - \omega_a + \frac{\Delta_a}{2} - \frac{1}{2}\sqrt{\Delta_a^2 + 4Ng^2}]^2}$$
$$+ sin^2\theta \frac{\frac{\Gamma_+}{\pi}}{\Gamma_+^2 + [\omega_p - \omega_a + \frac{\Delta_a}{2} - \frac{1}{2}\sqrt{\Delta_a^2 + 4Ng^2}]^2}. \quad (3.25)$$

The absorption spectrum of atoms is equal to the $Im\mathcal{X}(\omega_p)$ and follows the scaling-law for the atomic number $g \rightarrow g\sqrt{N}$. It can be observed that the imaginary part of that susceptibility contains a doublet type structure having peaks at $\omega_p = \omega_a + \Delta_a/2 \pm \sqrt{\Delta_a^2 + 4Ng^2}/2$, whose amplitude depend on factors Δ_a and $g\sqrt{N}$. In far-off detuning case, the peaks will be centered at the atomic transition frequency ω_a, however, in the resonant case $\Delta_a = 0$, these peaks will be shifted to $\omega_a \pm g\sqrt{N}$. These peaks in absorption quadrature correspond to the splitting of Rabi oscillation due to the vacuum states of intra-cavity photons. The absorption spectrum with such peaks leads to the concept of electromagnetically induced transparency, which will be discussed in detail in next section, which owns a key role to design quantum memory elements in quantum information. This formalism has been use to explain classical and quantum chaos with the help of cavity-Rydberg systems [13]. Recently, it has been used to reduce the effects of dc and low-frequency electromagnetic fields in cavity systems [14]. The expression for susceptibility Eq. (3.25) not only studies the influence of photon vacuum state on the trapped atoms but also is applicable for multi-atomic systems and even in the case of intra-cavity photon number state, or non-vacuum states [15].

3.3.3 Blackbody Radiations

To further extend our discussion of Rydberg atoms in cavities, it is important to mention the atomic energy spectrum behavior during blackbody excitations. When atoms are placed inside a cavity, their interactions with intra-cavity photon field, including vacuum state $|0\rangle$ and photon number states $|n\rangle$, are modified because of the boundary condition imposed by the cavity side mirrors. Like intra-cavity vacuum states effects, blackbody radiation effects will also be changed by the confinement of the cavity which lead to the alteration in absorption and emission spectrum of the atoms inside cavity. Therefore, in order to keep the evaluation of atomic system between cavity mirror near absolute zero temperature $T = 0K$, it is necessary to measure these blackbody radiation effects. However, in ground state atoms, blackbody radiation effects are less than the influence of intra-cavity vacuum states which appear in the form of shift in low-lying energy levels, ranging from kHz to few MHz [16–18]. While, in the case of Rydberg atoms, the shift in the atomic energy levels became smaller due to their highly excited states but the blackbody radiation become more importance in this situation because of blackbody excitations of low-lying energy levels which may create a problem in achieving Rydberg states in an atom.

Therefore, it's necessary to consider blackbody excitation effects in cavity-Rydberg system due to their large principal quantum number which amplify the influences of external heat bath in the form of blackbody excitations. The blackbody radiation effects on Rydberg atoms in free space has been discussed for both theoretical as well as experimental context [19–21]. But for cavity-Rydberg system, the blackbody thermal excitation was first theoretically investigated by Nakajima et al. [22]. They considered Rydberg atoms inside two parallel metallic plates separated by the distance equal to the wavelength of intra-cavity electromagnetic field which is supposed to be equal to the inter-atomic distance, the distance between two Rydberg dipoles. They calculated energy level sift due to the blackbody thermal excitation for the $24s_{1/2} \rightarrow 23p_{3/2}$ pair of states corresponding to the Rb Rydberg ensemble placed inside cavity. The black body radiation photons can influence the energy level configuration in two ways. First, they can modify the shift in low-lying energy level caused by the intra-cavity vacuum states. Second is, they may directly interact with the population of Rydberg atoms causing depopulation due to the absorption and emission of the blackbody radiation photons. In both ways, they will modify the energy level configuration of atomic states which is very crucial in the case of Rydberg atoms. The thermal shift due to blackbody excitation as a function of cavity length L and temperature of the external heat bath T reads as,

$$\Delta_{BB}(T, L) = -\frac{\alpha a_0^2 k_0 T}{3\hbar L^2} \sum_{j \neq i} d_{ji}^2 \sum_{n=0}^{\infty} P \int_n^{\infty} dk_0 k_0 \Big[\frac{1}{k_0 - 2L(E_i^w - E_j^w)} - \frac{1}{k_0 + 2L(E_i^w - E_j^w)} \Big] \times \Big[1 - (\frac{n}{k_0})^2 cos(\frac{2n\pi}{L} z) \Big], \quad (3.26)$$

where k_B is the Boltzmann constant, T is the temperature of external heat reservoir and d_{ji} represents electric dipole matrix elements. The parameter $k_0 = (L/\pi)k$, with cavity length L and incident wave vector k, is the effective dimensionless wave vector along the cavity axis and E_{ij}^{w} are the electromagnetic wave components for the atomic pair states i and j. n is the principal quantum number and α is used to represent single photon field with z spatial component long the cavity axis. It can be seen that the thermal shift in atomic energy spectrum is dependent on the temperature of reservoir, which is basically causing the injection of blackbody photon inside the cavity, and appears to be directly proportional to temperature T if the condition $\hbar c k_0 \pi / k_B T L << 1$ is satisfied.

They also illustrated the effects of blackbody excitation on the broadening of cavity mode spectrum and develop a relation for the spatial distribution of intra-cavity optical mode, in Fourier domain, as a function of cavity Q factor,

$$g(\omega) = \frac{\omega_c}{2\pi Q}\left[\frac{1}{(\omega - \omega_C)^2 + (\omega_c/2Q)^2}\right]. \tag{3.27}$$

It is interesting to note that if we consider the cavity with quality factor $Q \rightarrow \infty$, then the distribution of intra-cavity mode along the cavity is replaced by the delta function defined as, $g(\omega) = \delta(\omega - \omega_c)$. By using this cavity mode distribution relation in Eq. (3.26) and replacing intra-cavity field strength with quantized operators with weight $n\pi/L$, we can calculate the resultant thermal shift in atomic low-lying energy levels, by blackbody thermal radiations, as a function of cavity potential line broadening [22].

Soon after this, the blackbody radiation effects on cavity-Rydberg system was experimentally demonstrated by Raimond et al. [23] and Lai et al. [24]. They investigated cavity mediated control over the blackbody thermal excitation of Rydberg atoms with transitional pair of states $13s_{1/2} \rightarrow 13p_{3/2}$ and $14s_{1/2} \rightarrow 14p_{3/2}$, trapped inside a gold illuminated cavity. In free space configuration, the transitional rate γ_0 from $s \rightarrow p$ pair of state depends on the blackbody radiation wavelength λ and as the s states of the Rydberg atoms is isotropic, therefore, one-third of the blackbody excitations are absorbed by the three-dimensional polarizations inside the cavity. However, in parallel plate configuration, separated by the distance L, with same strength of external thermal blackbody radiations, this $s \rightarrow p$ transitional rate will be modified due to the implication of cavity boundary conditions. They demonstrated the dependence of population transfer due to blackbody thermal excitation on the cavity length, as shown in Fig. 3.3a, where the dotted line represents ionization signal obtained from the Rydberg atoms while curve a, b and c represent known spontaneous emission for Rydberg states $13s$, ionization of the atoms below $14s$ and for all atoms below Rydberg state $14s$. In their experiment, the observed ionization signal from Rydberg atoms decay more than as predicted due to the addition of blackbody radiations depending upon the cavity length, which transfer atoms from the low-lying states to the high-lying states with longer lifetime.

They count the ionization strength of atomic beam, by considering the atomic distribution along the cavity axis to measure the elapsed time as a distance, at room

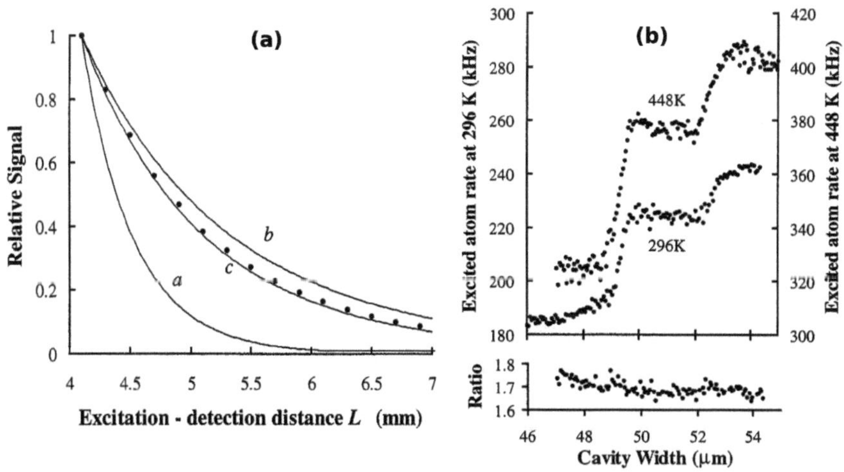

Fig. 3.3 **a** The excitation of Rydberg atoms inside cavity obtained from relative ionization signal, as a function of cavity length L. Curves a, b and c indicate the prediction of spontaneous emission for the state $13s$, high-lying state to $14s$ and $13s$, respectively. Dotted curve represents experimental observation of spontaneous emission for the Rydberg state $13s$ which is incompatible to the predicted results due to the addition of blackbody thermal excitations. In **b**, the upper panel shows the thermal excitation of Rydberg state $14s$ as a function of cavity length L for two different temperatures $T = 448$K and 296K. The lower panel illustrates the ratio of these two, temperature dependent, excitation signals which appears to be constant near 1.64. Adopted from [24], ©(2025) with permission from American Physical Society

temperature in order to see the effects of blackbody thermal excitations. It can be observed that the blackbody ionization signal show linear behavior revealing that the blackbody excitation lower than the cavity mediated atomic decay rate. But considering their influence on Rydberg state $13s$ and $13p$, it is very necessary to take care of these thermal excitations in order to sustain Rydberg states and their long-range interactions. In order to further illustrate the effects of blackbody thermal excitations on cavity-Rydberg system, they measure the excitation signal for $14s$ state Rydberg atoms at higher temperature, as illustrated in Fig. 3.3b, where the temperature of external heat bath is increased to $T = 296K$ and $= 448K$ at constant pump laser and atomic beam intensities to precisely measure the influence of blackbody radiations. It can be noted that the ionization of the intra-cavity atoms has increased with an increase in external temperature which cause the addition of blackbody photons depending upon the cavity length. Intuitively speaking, if the distance between the mirrors is larger, the cavity will accommodate more blackbody photons. The transformation rate for thermal excitations of the cavity-Rydberg system due to blackbody photons can be defined as $[exp(hc/\lambda kT) - 1]^{-1}$. This relation is very important in order to implement good-cavity configurations.

In another experiment, the blackbody induced single-photon detection has been performed by considering low-temperature cavity-Rydberg system [25]. They used selective-field-ionization method to count and detect a blackbody excited photon

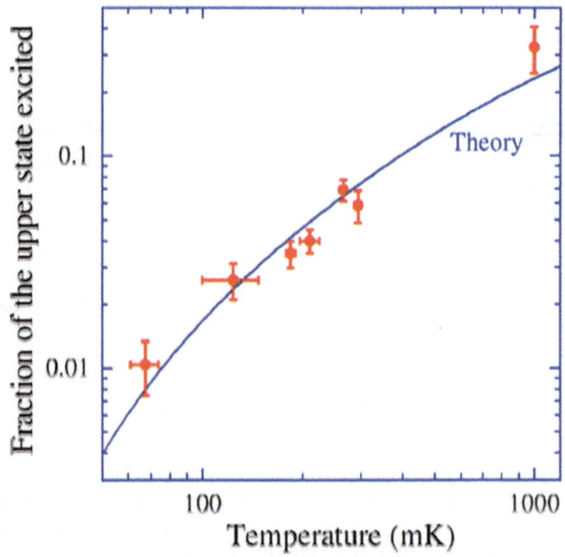

Fig. 3.4 The blackbody radiation effects on the fractional of Rydberg state $111p_{3/2}$ excitation as a function of temperature of the external heat bath. The red points correspond to the experimental results and blue cure is for theoretical calculation. Adopted from [25], ©(2025) with permission from Elsevier

from $111p_{3/2}$ excited state of Rydberg atom, which were initially $111p_{1/2}$, in cavity. Figure 3.4 shows, both theoretically and experimentally, the excitation of fraction of Rydberg atoms with increase in temperature of external heat bath. The increase in excitation with the increase in temperature can be observed which is measured by a single photon detector. The theoretical results are obtained by computing time-dependent evolution of density matrix of cavity-Rydberg system in interaction with external heat bath. Further, blackbody effects have been studied in Rydberg-Rydberg transitions in Coplanar microwave waveguide [26] and recently, the suppression of blackbody thermal excitations have been observed in cold Rydberg atoms [27].

3.4 Photon Fock State

In cavity-Rydberg systems, the preparation and evolution of intra-cavity photon number states or photon Fock states, a quantum state with well-defined number of photons lying in Fock space, are also particularly important in order to visualize whole system. Because of the presence of a particular number of discrete photons, these states are independent of external and intra-cavity electromagnetic intensity fluctuations and have indeterminate phase. Their Wigner representation contains negative probabilities which reveal that these states cannot be described by the classical picture and have particular quantum nature [28]. The negative regions of the Wigner function are direct signatures of non-classicality, providing a useful tool for identifying and quantifying quantum coherence and entanglement in light-matter interactions. The extensive investigations have been done in this regard to prepare and characterize

Fock states. It is comparatively easy to obtain single-photon number states because of the entanglement between the pair of photons which allows to predict the state of second entangled photon by simply measuring the state of first entangled photon [28, 29]. These single-photon states can also be engineered by controlling the radiation emission spectrum of highly excited molecules [30] and quantum dots [31]. Particularly in quantum dot systems, the emission statistics can be tailored using Purcell enhancement and resonant pumping schemes, which offer deterministic single-photon generation with sub-Poissonian statistics and strong photon antibunching.

These photon Fock states have contributed a significant role in developing the foundation of quantum information [32]. They serve as primary resources for quantum communication, quantum cryptography, and quantum computation by facilitating tasks such as teleportation, superdense coding, and quantum error correction. In cavity environment, these Fock states are obtained by observing interaction between intra-cavity optical field with atoms passing through high-Q cavity. The strong atom-photon coupling in these cavities ensures that the photonic state evolves in a predictable manner due to the well-defined interaction time and quantized field structure. The single-photon Fock states have been engineered in such cavities by manipulating quantum Rabi oscillation in π configuration [33] and adiabatically arranging atoms passing through the cavity [34]. In π-pulse configuration, a resonant atom-field interaction for a specific time duration drives the atom from excited to ground state while transferring one photon to the cavity, thus enabling precise control over the cavity photon number. Meanwhile, adiabatic schemes employ time-dependent variation of coupling strengths and detunings, ensuring robustness against decoherence and systematic errors.

In order to obtain multiple photon number states, a beam of highly excited atoms passing through the high-Q cavity, where each atom radiates single photon, provides a feasible setup. This method has been thoroughly explained and used in theoretical preparation of two-photon Fock states in Ref. [35], where two-photon Fock state was engineered by using two-atom laser source. Here, synchronization and spatial coherence of the atomic beam play a vital role in ensuring constructive quantum interference and emission of multiple photons into a single cavity mode. Two-photon Fock states have also been experimentally demonstrated by using highly excited Rydberg atoms in an optical cavity [36]. The Rydberg atoms, due to their exaggerated electromagnetic properties, large dipole moments, and long radiative lifetimes, serve as ideal sources for precise quantum-state engineering of intra-cavity fields. These experiments marked an essential step towards quantum state synthesis and validation of quantum electrodynamics (QED) predictions in controlled environments. Such photon-number-resolved state preparation and detection provide not only a testbed for fundamental quantum mechanics but also open avenues for the realization of deterministic quantum light sources and photon-based quantum gates essential for quantum networking and scalable quantum processors.

In the cavity-Rydberg system, atoms in the excited state are injected into high finesse cavity which decays to the ground state after emitting photons to the inter-cavity field. The interaction of atom and intra-cavity optical field, having vacuum state $|0\rangle$ initially, changes the state of cavity field in the form of superposition of number states $|n\rangle$ and $|n+1\rangle$ because of the addition of a radiated photon. Precisely saying, if the atom inside the cavity is measured in ground state than the intra-cavity number state will be in the order of $|n+1\rangle$ and if not, the intra-cavity photon number state will remain the same $|n\rangle$. Therefore, by detecting the state of an atom inside the cavity, we can reveal the pure number state of the cavity field instead of superposition. Here it should be noted that if the intra-cavity field is dissipative because of having low Q factor, it is very difficult to measure the precise number state due to the cavity leakage. If the intra-cavity atoms interact with photon number state $|n\rangle$ then the relative population exchange between ground state and excited state will occur at the rate of $-\sqrt{n+1}$, which reveals the intra-cavity population inversion as, $I = p_g - p_e$, where p_g and p_e are the probability amplitudes for finding atoms in ground state and excited state, respectively. When the intra-cavity field is not in a number state, which means dissipative cavity, then the population inversion will become [36],

$$I(t_{in}) = -C\sum_{n} P_n cos(2-\sqrt{n+1}t_{in}), \tag{3.28}$$

where t_{in} is the interaction time of the intra-cavity field with atoms and P_n represents the probability of finding the number n of intra-cavity photons. The parameter C accumulates the reduction of the cavity field because of the cavity mediated dark-state excitations. By using this relation, one can determine the distribution of the intra-cavity photon number by manipulating the Rabi transitional oscillations of the intra-cavity atoms.

In experiment [36], they measure the population inversion in intra-cavity atoms with respect to the atom-field interaction time and predict the probability of finding intra-cavity number state, as show in Fig. 3.5. Figure 3.5a–c, illustrate Rabi oscillation, computed from population inversion, as a function of interaction time corresponding to the photon number state 0, 1 and 2, respectively, while, in their comparison, Fig. 3.5d–f demonstrate the detection of probability amplitude corresponding to the photon number states. As all the atoms are in excited state initially, therefore, the Rabi transition will appear as phase independent. It can be noted that the probability amplitude for the targeted number states, around which the Rabi oscillations are considered, is maximum which provides proof for Fock states with $n = 1$ and 2. The factors that limit the chances of finding Fock states are the intra-cavity optical frequency and detuning fluctuations which they measure $3 \rightarrow 5\%$ in their experiment. However, these factors cannot affect the creation of Fock states itself because, inside the cavity, photon number states are only dependent on the phase of the Rabi oscillations. Figure 3.5g shows the dependence of photon number state on the Rabi oscillations in contrast to the Fig. 3.5d–f and presents the chances for all possible number state. After this, in cavity-Rydberg system, the multiple photon number states have been investigated by using Ramsey separated-oscillatory-fields method to detect atomic wave function phase shift [37], with the help of single-atom [38] and by using one-atom maser concept [39, 40]. Further, the controlled hole burning in intra-cavity photon number states, with the help of resonant interaction of intra-cavity

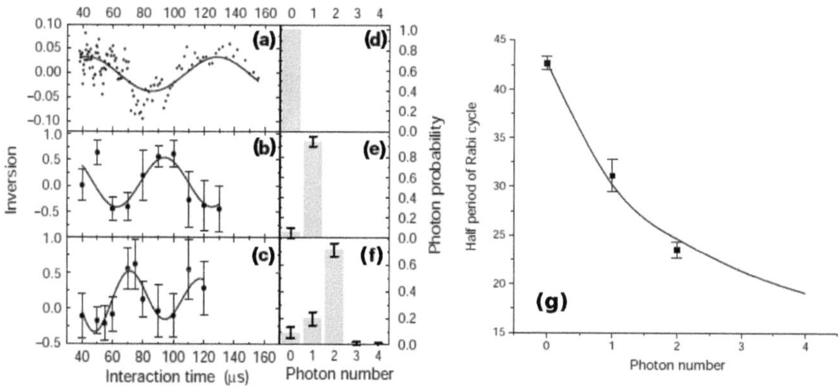

Fig. 3.5 **a–c** The Rabi oscillations as a function of interaction time between intra-cavity field and Rydberg atoms for intra-cavity photon number state $n = 0$, 1 and 2, respectively. **d–f** The probability distributions of photon number state P_n calculated from Eq. (3.28). In each plot, $P - n$ is calculated for number states from $n = 0 \rightarrow 3$. **f** The dependence of all possible intra-cavity photon number states on the Rabi oscillation after $\pi/2$ Rabi cycle in accordance to the results (**d**)–(**f**). Adopted from [36], ©(2025) with permission from Springer Nature

field and Rydberg atom, [41] and intra-cavity Fock state lifetime measurement with nondemolition photon count inside the cavity have been demonstrated [42].

In a recent experiment, the group of S. Haroche have been able to demonstrate and protect seven-photon number states in a cavity-Rydberg system by using circular Rydberg atoms [43]. For dispersive cavity, these Rydberg atoms possess quantum nondemolition behavior which means they can absorb and emit single photon to the cavity after interacting with intra-cavity resonant field. The schematic diagram for their experiment is shown in Fig. 3.6a, where C is the superconducting Fabry-Pérot cavity, with average blackbody thermal photons $n_{th} = 0.05$ and driven by the voltage source V, accompanied by two low-Q cavities R_1 and R_2 driven by two laser sources S_1 and S_2, respectively, generating pulses to manipulate atomic transitions. The circular $|g\rangle$ states of Rydberg atoms, with principal quantum number 50, are prepared in B with the help of pulsed laser with time period $T_a = 82\mu s$. The atomic beam enters into the cavity with velocity $v = 250$m/s and interacts with intra-cavity microwave field oscillating at $\omega_c = 51$GHz. The final state of atomic beam coming out of cavity is measured by the binary detector D and all the measurements are compiled through processor K. This three-cavity configuration is an analog of the Ramsey interferometer which measures quantum nondemolition behavior of Rydberg atomic states by detecting dispersive phase shift of cavity mode after interacting with circular atoms. They confirm the occurrence of Fock states by computing average probability distribution $p_t(n)$ independent to the controller K and match to the reference Poisson distribution which is constructed for the n_t photon number as shown in Fig. 3.6b. After $t = 140$ms, they interrupted the feedback loop and sent a beam of circular Rydberg atoms to the cavity C with Ramsey phase ϕ_r set to the eight Fock state con-

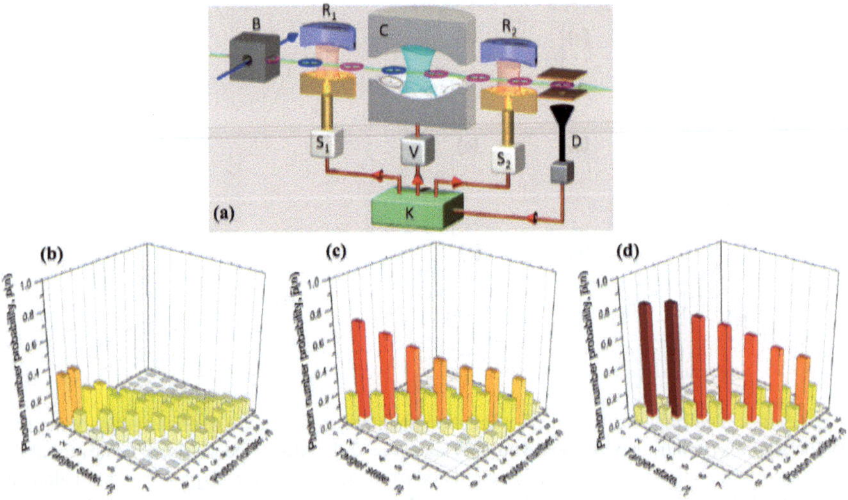

Fig. 3.6 **a** Schematic diagram for the experiment [43] performed for photon number states in cavity-Rydberg system composed of a superconducting cavity C surrounded by two low-Q cavities driven by two pulse laser sources S_1 and S_2. The voltage V is applied across the cavity C to produce electromagnetic field inside cavity. A beam of circular Rydberg atoms is produced in chamber B and detected via binary detector D. The controller K is used to process the data. **b** and **c** The histogram for intra-cavity photon number state probability distribution versus photon target state n_t and photon number n. For expected measurement (or without the interruption of K), the photon number states probability distribution is average over the 4000 events. However, the measurement for Fock type photon states probability is directly obtained by the detector K. Adopted from [43], ©(2025) with permission from American Physical Society

figuration [44]. On the average of 4000 events, the probability distribution for Fock states were plotted between photon number n and $n + 7$, as illustrated in Fig. 3.6c, by following the same procedure as demonstrated in [42]. The obtained distribution was appeared to be sub-Poissonian to the reference distribution $p_t(n)$ which represents the fidelities of Fock states corresponding to the target state. Further, to get better estimation, they involve the controller K to interrupt the feedback and measure the distribution for intra-cavity photon number states. The histogram shown in Fig. 3.6d illustrates the probability distribution for Fock states averaged over same number of events 4000 when K interrupts the feedback at $p_t(n) > 0.8$. It can be noted that the obtained fidelities for the initial three Fock states are very close to the expected value 0.8. For the third photon state, the 63% of the distribution reach the threshold in $t = 27$ms. The other number states are suppressed by the mismatch between classical and quantum nature of the sensors. However, still these states possess significant importance and can be stabilized as discussed in Ref. [45–47]. The creation and manipulation of these Fock states provide a platform to demonstrate decoherence and non-local dynamical of quantum physics. Further, these cavity-Rydberg photon number states are the direct implication of monochromatic micromasers [48], as discussed in the following section.

3.5 Rydberg Maser

A maser (microwave amplification by stimulation emission of radiation) is a type of laser that generates monochromatic amplified light from atoms by stimulated emission in visible light or ultraviolet [49–53]. In the case of Rydberg atoms in cavity, due to their large dipole matrix elements, Rydberg atoms interact strongly with infrared or microwave electromagnetic field which is resonant with the Rabi transitions between nearby energy levels revealing a scenario where coherent emission of light can be achieved in ultraviolet range. This is the main reason behind the strong emission of radioactive radiation in the interstellar median by Rydberg states [54]. The experimental demonstration of atomic maser in cavity-Rydberg system was made by Gross et al. in 1979 [55]. They detect maser effects in alkali Rydberg atoms interacting millimeter wavelength electromagnetic field inside cavity, driven by a dye laser, by measuring the population inversion in Rydberg states from intra-cavity field ionization. The cavity-Rydberg masers are much different from the previous maser devises due to the huge dipole matrix elements of Rydberg atoms, which are typically in order of 10^3 with principal quantum number up to 30. As a consequence of very low threshold, Rydberg atoms generate very low-power maser, up to 10^{-3}ev for each photon from each atom yielding in collective maser pulse of 1ev in their experiment. They detect cavity-Rydberg maser effects by injecting beam of excited atoms with low density, up to 10^4 excited atoms for each pulse, and by detecting output single after time delay $t_0 = 30\mu s$ corresponding to the total atomic time of flight inside cavity, as shown in Fig. 3.7. The ionization signal for Rydberg state $27s$ has been shown when the intra-cavity field is resonant (curve a) and slightly off-resonant (curve b) with Rabi transition frequency between the pair of state $27s \rightarrow 26p_{1/2}$. In this experiment wavelength of the radioactive emission is of the order of 1.49mm having energies about 10ev. One can clearly note the enhancement in the population inversion in low-lying state of $26s_{1/2}$ which in consequence reveals the maser effects. Here it should be noted that these results are still limited by the intra-cavity black-body radiation background, which cause the thermal excitation, and because of the spontaneous emission of radiation which affects the monochromatic nature of the masers. The length of the cavity also enables us to tune the maser excitation. If we swept the length of cavity, the maser effects will be shifted in position corresponding to the relation $\delta L = \lambda/2$, where length L is the length of the cavity and λ is the emission wavelength.

The generation of masers in cavity-Rydberg system was further extended by the same research ground in a experimented in which they developed heterodyne receiver for the detection of astronomical radio waves by illustrating emission of Rydberg

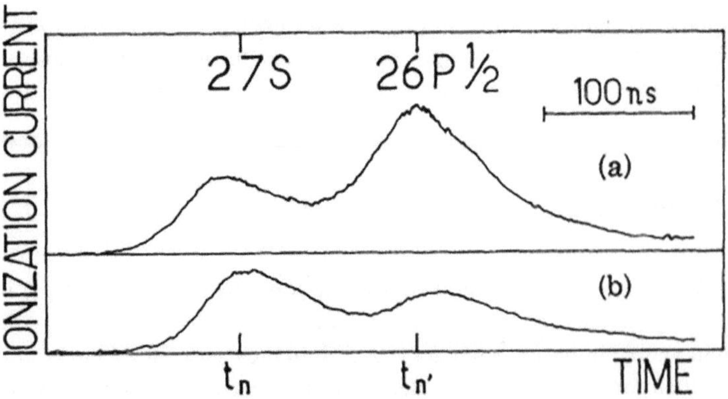

Fig. 3.7 The maser effects obtained from the ionization signal from the intra-cavity Rydberg Na atoms. The experimental readings are averaged over the 200 pulses. The curve 'a' accommodates ionization signal when intra-cavity field is resonant with the transition between Rydberg pair $27s \rightarrow 26p_{1/2}$ while curve 'b' represents 40MHz off-resonant case. The time t_n and $t_{n'}$ correspond to the two different peaks appearing in the results. Reused from [55], ©(2025) with permission from American Physical Society

atoms at microwave level [56]. In another investigation, they demonstrated, both theoretically as well as experimentally, the background blackbody radiation effects on microwave maser generated by cavity-Rydberg system [57]. Such masers are usually triggered by the intra-cavity blackbody background, which can be studied by adding a small wavelength signal to the cavity as a background. If we consider the intra-cavity optical mode resonant with atomic transition frequency and, for simplicity, at $T = 0$K, the periodic energy exchange between Rydberg atoms and intra-cavity field will occur at the rate intra-cavity Rabi oscillations:

$$g_a = \frac{d}{\hbar}\sqrt{\frac{2\hbar\omega_a}{\epsilon_0 \pi L \omega_0^2}}, \tag{3.29}$$

where parameter d is the electric dipole moment of the atom corresponding to the single intra-cavity photon. This oscillation rate should be in the order of inverse cavity damping time $T_{cav}^{-1} = \pi c/fL$ and if $g_0 > T_{cav}^{-1}$, the a number of Rabi oscillation can occur even before the cavity starts. On the other hand, if $g_0 < T_{cav}^{-1}$, cavity-Rydberg system will be over damped due to the intra-cavity atomic energies decay into radiations. In this situation, intra-cavity atomic energy loss will be in terms of $g_a^2 T_{cav} = f\Gamma Q$, where Γ is the intra-cavity photon decay rate, which indeed corresponds to the damped dynamics of cavity-Rydberg masers at rate T_R^{-1} with $N_0 = 1$ initial number of atoms. Such damped dynamics can only be observed if the intra-cavity decay rate is larger than the atomic escape time from cavity T_t^* which reveals the modified condition as, $f > 1/(\Gamma Q T_t^*)$. These conditions are illustrating the dependence of oscillatory maser process on the Q factor of the cavity. Further, the condition for atomic decay rate is appeared to be an analog of the threshold condition for Rydberg excitations, given as,

$$T_R ln[\frac{N_0}{1+n_B}] \leq T_t^*, \tag{3.30}$$

which shows that until the threshold is achieved, the single-atom masers can practically operate in oscillating regime. Such oscillatory dynamics can simply be explain by quantum mechanics by modeling whole cavity-Rydberg system as a two-level system in which the state $|nS_{1/2}; 0\rangle$ represents an excited Rydberg atom with no photon in cavity and state $|n \pm 1P_{1/2}; 1\rangle$ corresponds to the situation when atom decays to the lower energy level by emitting a photon to the cavity. The coupling between these states can be defined in terms of Rabi oscillation between these levels. These advancements have provided the foundation for the study and implementation of a single-atom [39, 58–63] and a single-photon [51, 64, 65] micro-maser which are crucially important for achieving and sustaining squeezed states [65–68] of light in modern applications of quantum computing and quantum information [69].

The maser process has been observed by demonstrating photon exchange between a single Rydberg atom and a single mode of the cavity, having quality factor $Q = 8 \times 10^8$, at temperature $T = 2.0$K [39]. The temperature of the cavity crucially depend on the Q factor of the cavity. If the value Q factor is higher, there are less thermal excitations due to the external heat bath. In their experiment, they illustrated that the effective temperature can vary from $T = 4.3$K to $T = 2$K depending upon Q factor $Q = 1.7 \times 10^7$ to $Q = 8 \times 10^8$, respectively. The average number of blackbody photon was considered $n_{bl} = 1.5$ for temperature 2K. They chose ionization method to determine the Rydberg states corresponding to the different principal quantum number. Further, they observe the ionization signal and count the reduction of electrons during transition of Rydberg states from $63p$ to $61d$ as a function of cavity mode frequency in order to detect maser process, as shown in Fig. 3.8a. To obtain the transition from $63p_{3/2}$ to $61d_{3/2}$ Rydberg states, cavity was tuned by changing the applied voltage and ionization signal was measured at the same time. In Fig. 3.8a, the reduction in ionization signal from $63p_{3/2}$ can be observed around flux 800atoms/s, which is occurring at cavity mode frequency $\approx$ 21GHz, revealing the maser operations. Further, with the increase in atomic flux and intra-cavity field frequency, a small shift appears in the ionization signal, which is basically an AC-Stark shift, due to the virtual transitions of the Rydberg state $63p_{3/2}$ just after 50MHz of the maser process. The ac Stark shift can be interpreted as the multiple photon exchange during the interaction between cavity mode and Rydberg atoms. In another experiment [60], performed by the same group, the maser operation was observed at cavity temperature $T = 0.5$K by similarly observing the transition between Rydberg stats $63p_{3/2}$ and $61d_{3/2}$, see Fig. 3.8b. They decrease the cavity temperature by just increasing the Q factor of the cavity to $Q = 3 \times 10^{10}$. The electron reduction in ionization signal can be seen depending upon the density of atoms inside cavity which reveals that the maser signal can be detected even the average number of intra-cavity

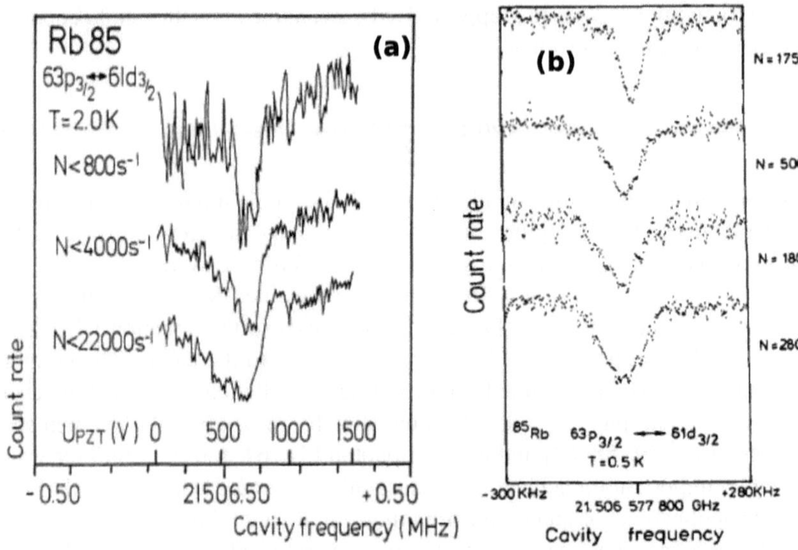

Fig. 3.8 The electron count from the ionization signal obtained during the Rydberg state transition from $63p_{3/2}$ to $61d_{3/2}$ in ^{85}Rb atoms inside cavity. **a** The maser resonance in the form of reduction in ionization signal as a function of intra-cavity field frequency at temperature $T = 2$K and different number of atoms, as mentioned in plot. Adopted from [39], ©(2025) with permission from American Physical Society. **b** The maser resonance as a function of cavity frequency at temperature $T = 0.5$K for same Rydberg pair transition and different number of atoms. From [60], ©(2025) with permission from IOP Publishing

atoms is about 0.06. This concept of single-atoms masers possesses great interests in modern quantum measurement experiments.

After these advancements, the cavity-Rydberg masers have been investigated and manipulated in many directions of physics. These investigations include the multi-stable dynamics of masers in cavity-Rydberg system [70], nonlinearity of the coherently driven mesoscopic micromaser [71] and the recent demonstration of Rydberg-Rydberg state transition during the maser operation in Coplanar microwave waveguide [72]. Further, in another investigation the trapping of state corresponding to the cavity-Rydberg system, depending upon the intra-cavity photon number state, during the maser operation was observed [73]. Such state trapping can only be achieved after the evolution of intra-cavity atom-field interaction which means when the system is finally stabilized. They also demonstrated the stabilization of intra-cavity photon number state during the maser operations by using feedback mechanism. Furthermore, in [74], the cavity mediated dependence of the micromasers on the hyperfine structure of Rydberg atoms has been discussed by modeling cavity-Rydberg system as a three-level system. Figure 3.9 shows the maser resonance curve as a function of intra-cavity detuning at different number of excited states N_{ex}. The state $|b\rangle$ and $|c\rangle$ are assumed to be equally coupled to the third state $|a\rangle$ define the the Rabi transition between first two states as $g_{ab} = g_{ac} = 1$ while the Rabi transition corresponding to

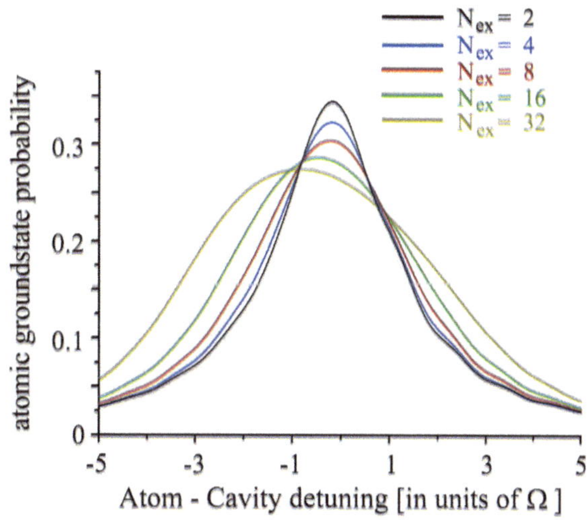

Fig. 3.9 The maser resonance curves obtained from ground state probability distribution as a function of intra-cavity optical field detuning for different excited state population of Rydberg atoms N_{ex}. Adopted from [74], ©(2025) with permission from Elsevier

the state $|a\rangle$ are in red detuning regime. If the probability of finding photon during the maser process matches the probability of finding atom in the state $|a\rangle$ $|c_a|^2$, maser resonance curves are achieved as shown in Fig. 3.9. It can be seen that if we decrease the number of excited states inside the cavity, the maser resonance curve is enhanced because of the purified maser photon excitation. The cavity-Rydberg masers have key importance in understanding many important applications of cavity-Rydberg systems.

3.6 Cavity Mediated Ultrafast Energy Exchange

The recent investigation on ultra-fast energy transfer between two Rydberg blockades [75]. The quantum optics platform combines several cutting-edge technologies to enable the study of Rydberg interactions at nanosecond timescales. Two ^{87}Rb atoms are laser-cooled to 5 μK in independent optical tweezers positioned with 10 nm stability using a feedback-stabilized optical system. The Rydberg excitation pathway employs a two-photon transition through the $5P_{3/2}$ intermediate state ($\Delta = 2\pi \times 200$ MHz detuning), achieving 90% population transfer in 30 ns pulses. Critical to the experiment is the electric field control system, which uses four gold-plated electrodes to create precisely adjustable fields ($0 - 5$ V/cm range) with sub-millivolt/cm stability, see Fig. 3.10. This allows exact tuning to the Förster resonance condition where the $|60S_{1/2}, 60S_{1/2}\rangle \leftrightarrow |60P_{1/2}, 59P_{3/2}\rangle$ energy matching occurs. State detection combines three techniques: (1) fluorescence imaging for initial atom positioning (0.5 μm resolution), (2) loss-free Rydberg state verification via shelving methods, and (3) time-resolved field ionization for joint state measurement. The entire system operates with 100 ps timing synchronization across all laser and detection components.

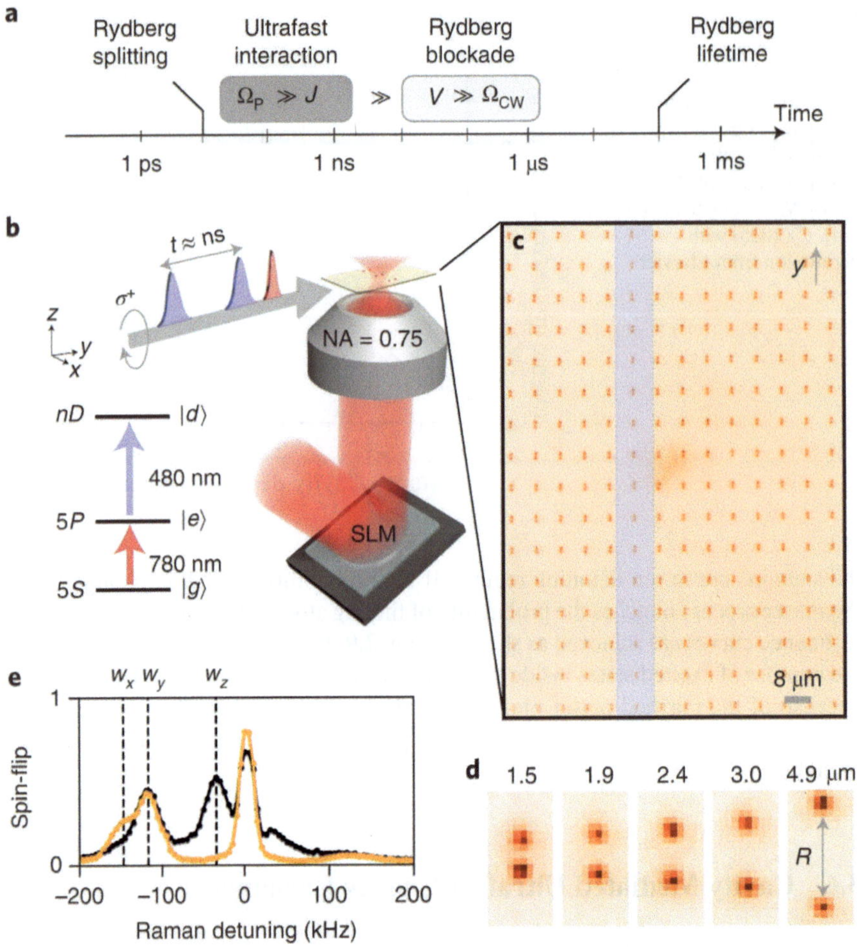

Fig. 3.10 **a** Ultrahigh vacuum chamber ($< 10^{-10}$ mbar) with anti-reflection coated viewports. **b** Dual optical tweezers (1064 nm, 1.5 μm waist, 0.7 mW power) generated by acousto-optic deflectors (AODs) with 10 nm positioning stability. **c** Rydberg excitation lasers: 780 nm (20 mW) and 480 nm (80 mW) with 2 kHz linewidth, focused to 3 μm spot size. **d** Electric field electrodes in tetrahedral configuration (2 mm spacing) with 16-bit DAC control (10 mV/cm resolution). **e** Detection system combining EMCCD imaging (90% quantum efficiency) and microchannel plate detector (50 ps timing resolution). Adopted from [75], licensed under CC-BY 4.0 from Springer Nature

This groundbreaking study [75] achieves three fundamental breakthroughs in coherent Rydberg interactions, with comprehensive experimental evidence presented in Fig. 3.11a–h. The central discovery demonstrates that when two individual rubidium-87 atoms are prepared in the $|60S_{1/2}, F = 2, m_F = 0\rangle$ state using a two-photon transition (780 nm + 480 nm lasers with <2 kHz linewidth) and positioned at a separation of 4.20 ± 0.05 μm in optical tweezers (1064 nm wavelength, 1.5 μm waist), they exhibit coherent energy exchange when subjected to a precisely controlled electric field of 3.12 ± 0.01 V/cm, see Fig. 3.11a. This condition creates an exact Förster resonance for the $|60S_{1/2}, 60S_{1/2}\rangle \leftrightarrow |60P_{1/2}, 59P_{3/2}\rangle$ transition, with the interaction strength following the characteristic dipole-dipole coupling $V_{dd} = C_3/R^3$ where $C_3 = 2\times\ 2.7 \pm 0.1$ GHz μm^3 was determined through careful calibration against DC Stark shift measurements.

The time-resolved dynamics in Fig. 3.11b reveal several remarkable features of this coherent energy exchange. The population oscillations between Rydberg states occur with a period of 8.5 ± 0.2 ns, corresponding to an interaction strength of 2× 118 ± 2 MHz, which represents approximately a 100-fold improvement in operation speed compared to state-of-the-art trapped ion systems. These oscillations maintain exceptional coherence properties, preserving 85 ± 2% contrast over 15 complete cycles (210 ± 5 ns total duration), despite the inherent decoherence mechanisms in Rydberg systems. Detailed analysis attributes the coherence decay to three primary factors: the natural lifetime of the 60S state (26 ± 1 ps, yielding $60S = 2\times\ 0.8$ MHz), electric field fluctuations (E = 10 ± 2 mV/cm causing $Stark = 2\times\ 4$ MHz inhomogeneous broadening), and residual thermal motion at the experimental temperature of 5 ± 0.5 K (resulting in = 0.03 rad/cycle phase variations). The resulting quality factor $Q = T_2\ 25$ represents a significant advancement for strongly interacting quantum systems.

The electric-field dependence of these interactions, shown in Fig. 3.11c, d, reveals an exceptionally tunable system with three distinct operational regimes. Near the Förster resonance at 3.12 V/cm, the interaction strength follows a sharply defined Lorentzian profile $V_{dd}(E) = \left(2\pi \times 150 \pm 3 \text{ MHz}\right) \Big/ \left[1 + \left(\frac{E - 3.12 \text{ V/cm}}{35 \pm 2 \text{ mV/cm}}\right)^2\right]$, allowing precise control over the interaction strength. In the resonant regime ($|E - E_{\text{res}}| < 35$ mV/cm), the system achieves maximum coupling strength with 85 ± 2% contrast in population oscillations, enabling the fastest quantum operations. The intermediate regime (35–105 mV/cm detuning) provides smoothly scalable interactions following $V_{dd} \propto 1/(E - E_{\text{res}})$, particularly valuable for analog quantum simulation of variable-range spin models. Complete suppression to $V_{dd} < 2\pi \times 1$ MHz occurs in the off-resonant regime (> 175 mV/cm detuning), allowing selective isolation of specific atom pairs in larger arrays. The system demonstrates rapid switching between these regimes in < 2 ns through optimized field ramping protocols, combined with exceptional stability (< 1% variation over 10^4 experimental cycles), as verified by Allan deviation measurements.

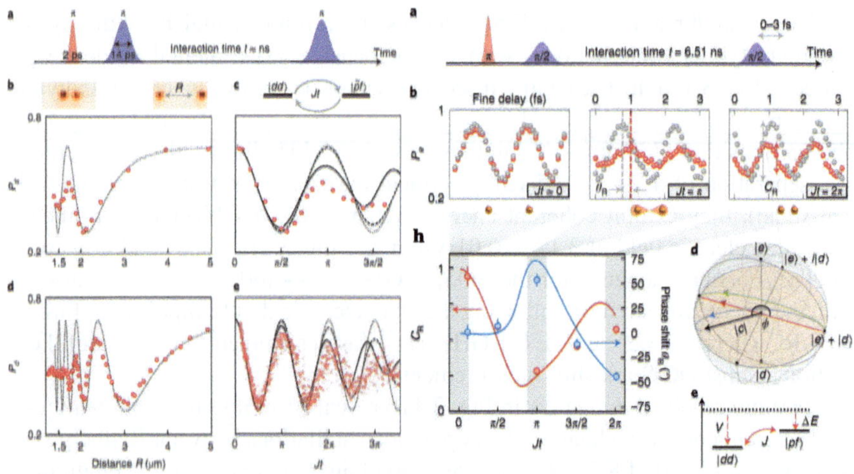

Fig. 3.11 Coherent energy exchange between two $|60S_{1/2}\rangle$ Rydberg atoms. **a** Experimental sequence showing off-resonant dressing lasers ($\lambda = 480$ nm, $\Delta = 2\pi \times 10$ MHz, $\Omega = 2\pi \times 3$ MHz) and electric field control. **b** Time-resolved population oscillations (period 8.5 ± 0.2 ns, $T_2 = 210 \pm 5$ ns) demonstrating $V_{dd} = 2\pi \times 118 \pm 2$ MHz coupling at 4.2 μm separation. **c** Electric field dependence revealing Förster resonance at 3.12 ± 0.01 V/cm with Lorentzian fit (FWHM=35 ± 2 mV/cm). **d** Interaction potential showing $C_3 = 2\pi \times 2.7 \pm 0.1$ GHz μm^3 from Stark shift calibration. Remaining show the quantum control applications. **e** Distance-dependent blockade showing R^{-6} to R^{-12} crossover at 4 μm and $R_c = 6.8 \pm 0.2$ μm blockade radius. **f** 1.6 ns blockade establishment time. **g** CNOT gate operation with $92.5 \pm 0.7\%$ fidelity at 5 μm spacing. **h** Entanglement generation at 1.25×10^7 Bell pairs/second. All error bars represent $\pm 1\sigma$ from 10^4 experimental repetitions. Obtained from [75], licensed under CC-BY 4.0 from Springer Nature

Figure 3.11 also presents systematic characterization of the Rydberg blockade effect under these strong interaction conditions. The distance-dependent excitation probability in Fig. 3.11e follows a composite scaling law $P_{\text{exc}}(R) = 0.25\left[1 - \exp\left(-\left(\frac{6.8\,\mu\text{m}}{R}\right)^6\right)\right] + 0.05\left(\frac{1\,\mu\text{m}}{R}\right)^{12}$, where the R^{-6} van der Waals behavior at large separations crosses over to an R^{-12} dependence below 4 μm due to higher-order multipole interactions becoming significant at short range. The extracted blockade radius of 6.8 ± 0.2 μm at $V_{dd} = 2\pi \times 100$ MHz agrees with theoretical predictions within measurement uncertainty, while the crossover to R^{-12} scaling provides new insights into the breakdown of simple dipole–dipole models at sub-micron distances. Temporal resolution of the blockade establishment in Fig. 3.5b shows complete suppression developing in just 1.6 ± 0.1 ns, representing a three-orders-of-magnitude improvement over previous demonstrations in thermal atom ensembles.

The quantum information processing capabilities are demonstrated through high-fidelity gate operations shown in Fig. 3.11f–h. Quantum process tomography of a CNOT gate implemented at 5 μm spacing reveals $92.5 \pm 0.7\%$ fidelity, verified through 10,000 experimental repetitions. The gate performance remains robust against technical imperfections, with position fluctuations ($\sigma_R = 150 \pm 20$ nm) contributing $3 \pm 0.5\%$ error, laser intensity variations ($\delta\Omega = 2 \pm 0.3\%$) causing $1.5 \pm 0.2\%$ pulse-area errors, and state preparation/measurement imperfections adding $2 \pm 0.5\%$ baseline infidelity. The system sustains entanglement generation rates of 1.25×10^7 Bell pairs per second when operated in pulsed mode, as quantified in Fig. 3.5e. Projections based on measured parameters indicate straightforward scaling to 10-atom quantum registers with expected gate fidelities above 90%, with clear pathways toward fault-tolerant operation through improved optical trapping geometries (targeting $\sigma_R < 50$ nm) and cryogenic operation to enhance Rydberg state lifetimes beyond 100 μs.

These breakthroughs are enabled by the innovative experimental apparatus shown in Fig. 3.10, which combines several cutting-edge technologies. The dual optical tweezer system (1064 nm wavelength, 1.5 μm waist, 0.7 mW power per trap) provides 10 nm position stability through active feedback stabilization of the acousto-optic deflectors used for trap positioning. Rydberg excitation is achieved through a two-photon transition (780 nm + 480 nm) with intermediate state detuning of 2×200 MHz, delivering 90% population transfer in 30 ns pulses. The tetrahedral electrode configuration (2 mm spacing) with 16-bit digital-to-analog conversion enables electric field control with 10 mV/cm resolution, critical for precise Stark tuning. State detection combines three complementary techniques: fluorescence imaging for initial atom positioning (0.5 μm resolution), loss-free Rydberg state verification via shelving methods, and time-resolved field ionization for joint state measurement with 50 ps timing resolution. The entire system operates with <100 ps timing synchronization across all laser and detection components.

The combination of results presented in Fig. 3.11 establishes neutral atom arrays as a leading platform for quantum information science, offering unique advantages in speed, scalability, and control. The demonstrated nanosecond-scale operations with microsecond-scale coherence times open new possibilities for studying non-equilibrium quantum dynamics and implementing error-corrected quantum computation. The technical advances in electric field control and Rydberg state manipulation provide a blueprint for developing large-scale quantum processors capable of solving classically intractable problems in quantum chemistry and materials science. Future directions include integration with photonic interconnects for distributed quantum computing and exploration of novel many-body phases in programmable quantum simulators.

3.7 Coupling Cold Rydberg Atoms with Cavities

The field of ultracold atomic physics has seen revolutionary advances in the past few decades, primarily driven by the ability to cool and trap neutral atoms to microkelvin and nanokelvin temperatures. These advances have opened up new frontiers for the exploration of strongly correlated quantum systems, quantum simulation, precision metrology, and quantum information processing. One particularly exciting development in this domain is the use of highly excited Rydberg states in cold atomic gases [76].

Rydberg atoms are neutral atoms excited to high principal quantum numbers $n \gg 1$, where their valence electrons orbit far from the nucleus. This results in a remarkable set of exaggerated atomic properties, including enormous dipole moments, long lifetimes, and extreme sensitivity to external electric and magnetic fields. These features make Rydberg atoms ideal platforms for inducing strong, tunable interactions over micron-scale distances far exceeding those of ground-state atoms. Combining the precise control offered by ultracold atomic ensembles with the strong interactions of Rydberg states has emerged as a powerful tool for engineering collective many-body states, probing quantum phase transitions, and realizing hybrid quantum systems. In this chapter, we focus on the methods for preparing and exciting Rydberg states in cold atomic gases and the motivation for coupling such systems to optical or microwave cavities for enhanced control and readout.

The preparation of cold atoms suitable for Rydberg excitation typically begins with a magneto-optical trap (MOT), where counter-propagating laser beams and a quadrupole magnetic field confine and cool atoms to temperatures on the order of 100–200 μK. Further sub-Doppler cooling and optical molasses can reduce the temperature below 50 μK. These cold atoms are then transferred into optical dipole traps or optical lattices, where they can be held for extended durations without significant heating or decoherence. The atomic density, spatial configuration, and trapping geometry can be controlled with high precision, setting the stage for coherent Rydberg excitation.

3.7.1 Two-Photon Rydberg Excitation

Rydberg states are usually excited via two-photon transitions due to the large energy gap between the ground and Rydberg levels. For rubidium, a common scheme uses a red laser at 780 nm to couple the ground state $|5S_{1/2}\rangle$ to the intermediate $|5P_{3/2}\rangle$, and a blue laser at 480 nm to couple to the desired Rydberg state $|nS_{1/2}\rangle$ or $|nD_{5/2}\rangle$. The effective Rabi frequency Ω_{eff} for the two-photon transition is given by

$$\Omega_{\text{eff}} = \frac{\Omega_1 \Omega_2}{2\Delta},$$

where Ω_1 and Ω_2 are the Rabi frequencies of the lower and upper transitions, and Δ is the detuning from the intermediate state.

To mitigate the effects of spontaneous emission from the intermediate state, the detuning Δ is chosen to be large compared to the natural linewidth Γ, placing the system in the so-called far-off-resonant excitation regime.

3.7.2 Limitations of Free-Space Rydberg Excitation

Despite the promising interaction properties of Rydberg atoms, experiments in free space face challenges in detection sensitivity, decoherence, and controllability of individual excitations. The coupling strength between a single Rydberg excitation and a free-space probe field is typically weak, requiring large ensembles to produce measurable optical signals. Moreover, the spatial inhomogeneity of laser beams and the broad linewidths of free-space transitions can limit the fidelity of quantum state preparation and measurement [77].

Incorporating a high-finesse optical or microwave cavity into the Rydberg excitation platform enables significant enhancements in light-matter interaction strength. The cavity mode enhances the effective interaction cross-section via the Purcell effect, and supports collective strong coupling for ensembles of atoms. The interaction of N atoms with the cavity field scales as $\sqrt{N}g$, where g is the single-atom cavity coupling strength. Coupling Rydberg atoms to a cavity facilitates:

- Enhanced sensitivity to single or few Rydberg excitations via cavity transmission or reflection spectra.
- Coherent exchange of excitations between light and atomic degrees of freedom, enabling polariton formation.
- Exploration of many-body cavity QED phenomena, such as photon blockade and quantum phase transitions.
- Non-destructive readout of Rydberg populations via dispersive cavity shifts.

3.7.3 Experimental Realizations

Several pioneering experiments have demonstrated the integration of cold atomic ensembles with optical cavities for Rydberg excitation. These setups typically use Fabry–Pérot cavities or optical ring cavities aligned with the atomic cloud. The cavity may support probe or control fields, or serve purely as a detection tool by monitoring its transmission spectrum. One such experiment involved a cloud of ultracold rubidium atoms trapped within the mode volume of an optical cavity and excited to Rydberg states using a two-photon ladder scheme. The cavity transmission spectrum exhibited vacuum Rabi splitting modified by Rydberg-induced nonlinearities, signaling the emergence of Rydberg polaritons.

More recent proposals extend this architecture to include coupling of Rydberg ensembles to superconducting microwave cavities or photonic crystal cavities, opening pathways toward hybrid quantum networks. The excitation of Rydberg states in cold atomic ensembles provides a unique platform for studying long-range interacting quantum systems with high precision and control. Coupling these systems to optical or microwave cavities not only improves detection capabilities but also introduces novel regimes of cavity QED with strong, nonlinear, and collective atomic interactions. This hybrid approach forms the basis for the study of quantum nonlinear optics, cavity-mediated entanglement, and scalable quantum information processing with Rydberg atoms [78, 79].

References

1. E.T. Jaynes, F.W. Cummings, Comparison of quantum and semiclassical radiation theories with application to the beam maser. Proc. IEEE **51**, 89 (1963)
2. G. Rempe, H. Walther, N. Klein, Observation of quantum collapse and revival in a one-atom maser. Phys. Rev. Lett. **58**, 353 (1987)
3. S. Stenholm, Quantum theory of electromagnetic fields interacting with atoms and molecules. Phys. Rep. **6**(1), 1–121 (1973)
4. E.M. Purcell, H.C. Torrey, R.V. Pound, Resonance absorption by nuclear magnetic moments in a solid. Phys. Rev. **69**, 681 (1946)
5. P. Goy, J.M. Raimond, M. Gross, S. Haroche, Observation of cavity-enhanced single-atom spontaneous emission. Phys. Rev. Lett. **50**, 1903 (1983)
6. Y. Kaluzny, P. Goy, M. Gross, J.M. Raimond, S. Haroche, Observation of self-induced Rabi oscillations in two-level atoms excited inside a resonant cavity: the ringing regime of superradiance. Phys. Rev. Lett. **51**, 1175 (1983)
7. A.V. Kozlovskii, Rabi oscillations in a two-level atom moving in an open resonator with spherical mirrors. Quant. Elect. **32**(1), 71 (2002)
8. Y. Yang, J. Xu, G. Li, H. Chen, Interactions of a two-level atom and a field with a time-varying frequency. Phys. Rev. A **69**, 053406 (2004)
9. T. Meunier et al., Entanglement and decoherence of N atoms and a mesoscopic field in a cavity. Phys. Rev. A **74**, 033802 (2006)
10. B.J. Dalton, B.M. Garraway, J. Jeffers, S.M. Barnett, Grassmann phase space theory and the Jaynes-Cummings model. Ann. Phys. **334**, 100–141 (2013)
11. G.S. Agarwal, Vacuum-field Rabi splittings in microwave absorption by Rydberg atoms in a cavity. Phys. Rev. Lett. **53**, 1732 (1984)
12. C. Cohen-Tannoudji, S. Reynaud, Simultaneous saturation of two atomic transitions sharing a common level. J. Phys. B **10**, 345 (1977)
13. S.V. Prants, Chaos, fractals, and atomic flights in cavities. JETP Lett. **75**, 12 (2002)
14. L.A. Jones, J.D. Carter, J.D.D. Martin, Rydberg atoms with a reduced sensitivity to dc and low-frequency electric fields. Phys. Rev. A **87**, 023423 (2013)
15. M. Brune et al., From lamb shift to light shifts: vacuum and subphoton cavity fields measured by atomic phase sensitive detection. Phys. Rev. Lett. **72**, 3339 (1994)
16. D.J. Heinzen, M.S. Feld, Vacuum radiative level shift and spontaneous-emission linewidth of an atom in an optical resonator. Phys. Rev. Lett. **59**, 2623 (1987)
17. V. Sandoghdar, C.I. Sukenik, E.A. Hinds, S. Haroche, Direct measurement of the van der Waals interaction between an atom and its images in a micron-sized cavity. Phys. Rev. Lett. **68**, 3432 (1992)
18. C.I. Sukenik, M.G. Boshier, D. Cho, V. Sandoghdar, E.A. Hinds, Measurement of the Casimir-Polder force. Phys. Rev. Lett. **70**, 560 (1993)
19. A.G. Vaidyanathan, W.P. Spencer, D. Kleppner, Observation of cavity-enhanced single-atom spontaneous emission. Phys. Rev. Lett. **47**, 1592 (1981)
20. D. Kleppner, Inhibited spontaneous emission. Phys. Rev. Lett. **47**, 233 (1981)
21. L. Hollberg, J.L. Hall, Measurement of the shift of Rydberg energy levels induced by blackbody radiation. Phys. Rev. Lett. **53**, 230 (1984)
22. T. Nakajima, P. Lambropoulos, H. Walther, Level shift and depopulation by blackbody radiation of a Rydberg atom between two metallic plates. Phys. Rev. A **56**, 5100 (1997)

23. J.M. Raimond et al., Collective absorption of blackbody radiation by Rydberg atoms in a cavity: an experiment on Bose statistics and Brownian motion. Phys. Rev. Lett. **49**, 117 (1982)
24. K.S. Lai, E.A. Hinds, Blackbody excitation of an atom controlled by a tunable cavity. Phys. Rev. Lett. **81**, 2671 (1998)
25. M. Tada et al., Single-photon detection of microwave blackbody radiations in a low-temperature resonant-cavity with high Rydberg atoms. Phys. Lett. A **349**, 488 (2006)
26. S.D. Hogan et al., Driving Rydberg–Rydberg transitions from a coplanar microwave waveguide. Phys. Rev. Lett. **108**, 063004 (2012)
27. T. Zhou, B.G. Richards, R.R. Jones, Absence of collective decay in a cold Rydberg gas. Phys. Rev. A **93**, 033407 (2016)
28. A.I. Lvovsky et al., Quantum state reconstruction of the single-photon Fock state. Phys. Rev. Lett. **87**, 050402 (2001)
29. P. Grangier, G. Roger, A. Aspect, Experimental evidence for a photon anticorrelation effect on a beam splitter: a new light on single-photon interferences. Europhys. Lett. **1**, 173 (1986)
30. Ch. Brunel et al., Triggered source of single photons based on controlled single molecule fluorescence. Phys. Rev. Lett. **83**, 2722 (1999)
31. S.R. Friberg, S. Machida, Y. Yamamoto, Quantum-nondemolition measurement of the photon number of an optical soliton. Phys. Rev. Lett. **69**, 3165 (1992)
32. E. Knill, R. Laflamme, G.J. Milburn, A scheme for efficient quantum computation with linear optics. Nature (London) **409**, 46 (2001)
33. X. Maître et al., Quantum memory with a single photon in a cavity. Phys. Rev. Lett. **79**, 769 (1997)
34. M. Hennrich et al., Vacuum-stimulated Raman scattering based on adiabatic passage in a high-finesse optical cavity. Phys. Rev. Lett. **85**, 4872 (2000)
35. H. Fearn, R. Loudon, Theory of two-photon interference. J. Opt. Soc. Am. B **6**, 917–927 (1989)
36. B.T.H. Varcoe et al., Preparing pure photon number states of the radiation field. Nature (London) **403**, 743 (2000)
37. M. Brune, S. Haroche, V. Lefevre, J.M. Raimond, N. Zagury, Quantum nondemolition measurement of small photon numbers by Rydberg-atom phase-sensitive detection. Phys. Rev. Lett. **65**, 976 (1990)
38. P. Bertet et al., Generating and probing a two-photon Fock state with a single atom in a cavity. Phys. Rev. Lett. **88**, 143601 (2002)
39. D. Meschede, H. Walther, G. Müller, One-atom maser. Phys. Rev. Lett. **54**, 551 (1985)
40. N. Nayak, D. Das, Theory of the Rydberg-atom one-photon micromaser. Phys. Rev. A **48**, 2475 (1993)
41. A.T. Avelar, B. Baseia, Controlled hole burning in Fock space via resonant interaction. Phys. Rev. A **72**, 025801 (2005)
42. M. Brune et al., Process tomography of field damping and measurement of Fock state lifetimes by quantum nondemolition photon counting in a cavity. Phys. Rev. Lett. **101**, 240402 (2008)
43. X. Zhou et al., Field locked to a Fock state by quantum feedback with single photon corrections. Phys. Rev. Lett. **108**, 243602 (2012)
44. C. Guerlin et al., Progressive field-state collapse and quantum non-demolition photon counting. Nature (London) **448**, 889 (2007)
45. C. Sayrin et al., Real-time quantum feedback prepares and stabilizes photon number states. Nature (London) **477**, 73 (2011)
46. I. Dotsenko et al., Quantum feedback by discrete quantum nondemolition measurements: towards on-demand generation of photon-number states. Phys. Rev. A **80**, 013805 (2009)
47. J. Audretsch, T. Konrad, A. Scherer, Quantum-optical weak measurements can visualize photon dynamics in real time. Phys. Rev. A **65**, 033814 (2002)
48. P. Maioli et al., Nondestructive Rydberg atom counting with mesoscopic fields in a cavity. Phys. Rev. Lett. **94**, 113601 (2005)
49. J.P. Gordon, H.J. Zeiger, C.H. Townes, The Maser-new type of microwave amplifier, frequency standard, and spectrometer. Phys. Rev. **99**, 1264 (1955)

50. B.T.H. Varcoe, S. Brattke, H. Walther, Generation of Fock states in the micromaser. J. Opt. B: Quantum Semiclass. Opt. **2**, 154 (2000)
51. K. Shimoda, T.C. Wang, C.H. Townes, Further aspects of the theory of the maser. Phys. Rev. **102**, 1308 (1956)
52. M. Tavis, F.W. Cummings, Approximate solutions for an N-molecule-radiation-field Hamiltonian. Phys. Rev. **188**, 692 (1969)
53. R. Bonifacio, G. Preparata, Coherent spontaneous emission. Phys. Rev. A **2**, 336 (1970)
54. B. Höglund, P.G. Mezger, Hydrogen emission line $n_{110} \to n_{109}$: detection at 5009 MH in galactic H II regions. Science **150**, 339 (1965)
55. M. Gross et al., Maser oscillation and microwave superradiance in small systems of Rydberg atoms. Phys. Rev. Lett. **43**, 343 (1979)
56. L. Moi et al., Heterodyne detection of Rydberg atom maser emission. Opt. Commun. **33**, 47 (1980)
57. L. Moi et al., Rydberg-atom masers I a theoretical and experimental study of super-radiant systems in the millimeter-wave domain. Phys. Rev. A **27**, 2043 (1983)
58. S. Singh, Quantum statistical theory of the ring laser. Phys. Rev. A **23**, 837 (1981)
59. H. Walther, The single atom maser and the quantum electrodynamics in a cavity. Phys. Scr. **T23**, 165–169 (1988)
60. G. Rempe, M.O. Scully, H. Walther, The one-atom maser and the generation of nonclassical light. Phys. Scr. **T34**, 5–13 (1991)
61. R.G. Hulet et al., Inhibited spontaneous emission by a Rydberg atom. Phys. Rev. Lett. **55**, 2137 (1985)
62. P. Filipowicz, J. Javanainen, P. Meystre, Theory of a microscopic maser. Phys. Rev. A **34**, 3077 (1986)
63. A. Datta et al., Information transfer through a one-atom micromaser. Europhys. Lett. **67**(6), 934–940 (2004)
64. N. Nayak, Influence of reservoir-induced interactions on the one-photon micromaser action. Opt. Commun. **118**, 114–122 (1995)
65. J. Krause et al., Preparation of a pure number state and measurement of the photon statistics in a high-Q micromaser. Phys. Rev. A **39**, 1915 (1989)
66. J. Krause, M.O. Scully, H. Walther, State reduction and $|n\rangle$-state preparation in a high-Q micromaser. Phys. Rev. A **36**, 4547(R) (1987)
67. L.A. Lugiato, M.O. Scully, H. Walther, Connection between microscopic and macroscopic maser theory. Phys. Rev. A **36**, 740 (1987)
68. A. Heidmann, J.M. Raimond, S. Reynaud, Squeezing in a Rydberg atom maser. Phys. Rev. Lett. **54**, 326 (1985)
69. M.P. Klembovsky et al., Quantum bit detector. JETP Lett. **79**, 441–444 (2004)
70. F. Casagrande, L.A. Lugiato, W. Lange, H. Walther, Multistability in mesoscopic Rydberg-atom systems. Phys. Rev. A **48**, 790 (1993)
71. C. Balconi, F. Casagrande, A. Mondini, W. Lange, Nonlinear dynamics of a passive, coherently driven mesoscopic micromaser. Phys. Rev. A **54**, 898 (1996)
72. S.D. Hogan et al., Driving Rydberg–Rydberg transitions from a coplanar microwave waveguide. Phys. Rev. Lett. **108**, 063004 (2012)
73. M. Weidinger, B.T.H. Varcoe, R. Heerlein, H. Walther, Trapping states in the micromaser. Phys. Rev. Lett. **82**, 3795 (1999)
74. P. Thoumany, T. Becker, Atomic hyperfine structure in micromaser physics. Opt. Commun. **285**, 2382–2386 (2012)
75. Y. Chew et al., Ultrafast energy exchange between two single Rydberg atoms on a nanosecond timescale. Nat. Photon. **16**, 724–729 (2022)
76. T.F. Gallagher, *Rydberg Atoms* (Cambridge University Press, 1994)
77. M. Saffman, T.G. Walker, K. Mølmer, Quantum information with Rydberg atoms. Rev. Mod. Phys. **82**, 2313 (2010)
78. O. Firstenberg, C.S. Adams, S. Hofferberth, Nonlinear quantum optics mediated by Rydberg interactions. J. Phys. B: At. Mol. Opt. Phys. **49**, 152003 (2016)
79. J.D. Pritchard et al., Cooperative atom-light interaction in a blockaded Rydberg ensemble. Phys. Rev. Lett. **105**, 193603 (2010)

Chapter 4
Nonlinear Optics with Blockade and Polaritons

Abstract This chapter discusses the emergence of strong optical nonlinearities from Rydberg interactions. Key effects such as the single-photon Kerr nonlinearity, electromagnetically induced transparency (EIT), and their quantum extensions are developed. The role of the cavity in enhancing blockade and enabling cavity-Rydberg polaritons is analyzed in depth. Particular attention is given to the formation of dark-state polaritons and their utility for realizing quantum nonlinear media, paving the way toward single-photon switches and quantum light sources.

4.1 Introduction

The stunning manifestation of quantum description of light lies behind the concept of information transfer in quantum communication and computation, where the light quanta (photon) is considered as best long-distance information carrier. Due to their low decoherence, high speed, and ease of transmission through fiber-optic networks or free space, photons serve as an ideal medium for carrying quantum information across large distances. However, the photons possess linearity by not directly interacting with each other which makes quantum information process difficult. This lack of natural photon-photon interaction prevents direct implementation of essential quantum logic operations, such as controlled-phase gates or entanglement swapping, which are otherwise straightforward in systems with inherent particle-particle interactions. Therefore, to fully manipulate the quantum characteristics of light, it is necessary to generate nonlinearity at single-photon level by exploiting the strength of their mutual interactions. Such nonlinear interactions can be triggered by manipulating photonic degree of freedom combined with other auxiliary mediums, such as cold and ultra-cold atoms. These atomic systems offer precise control over light-matter interactions due to their long coherence times and tunable internal energy levels.

These mediums can have either off-resonant two-level configuration or resonant three-level configuration with light, where the nonlinearity can be measured by the electromagnetically induced transparency (EIT) phenomenon. In the EIT framework,

K. A. Yasir and W.-M. Liu, *Rydberg Atoms in Cavity*, Springer Series in Solid-State Sciences 213, https://doi.org/10.1007/978-981-95-4571-1_4

a control laser field can render an otherwise opaque medium transparent to a probe field by inducing destructive interference between excitation pathways, resulting in reduced absorption and enhanced dispersion. This coherent phenomenon not only slows down light but also facilitates strong optical nonlinearities at low photon numbers, making it an essential tool for quantum optics and photonic quantum information processing. The interaction of optical degree of freedom with each other and with other physical objects gets further enhanced because of the confined environment induced by the high finesse cavity. The cavity boosts the field intensity and prolongs interaction time through multiple photon reflections, which effectively increases the nonlinear response and makes single-photon nonlinear effects experimentally accessible.

Moreover, the inclusion of Rydberg atoms provides another opportunity to further enhance such nonlinear optical interactions because of the long-range dipole-dipole interactions. Rydberg atoms, with their high principal quantum numbers, exhibit enormous polarizability and interaction radii, enabling interatomic interactions over micrometer distances orders of magnitude larger than typical atomic scales. Such long-range interaction can make Rydberg excitation off-resonant with intra-cavity field leading towards the so-called blockade phenomenon, where the whole Rydberg ensemble behaves as a single two-level system and can be perfectly explained by the JC Hamiltonian. This phenomenon suppresses the simultaneous excitation of more than one atom within a certain volume termed the blockade radius thus ensuring that a single excitation prevents others nearby, an effect that translates directly to photon blockade when mediated through cavity QED setups. When these blockade excitations are evaluated in EIT configuration, a significant amount of nonlinearity can be generated. The combination of EIT and Rydberg blockade leads to the formation of Rydberg polaritons, quasi-particles that embody both strong interactions and photonic coherence, which are capable of exhibiting giant Kerr nonlinearities even at the single-photon level.

Furthermore, these optical nonlinearities are the reason behind the blockade of photon number state in cavity and the creation of Rydberg polariton state. The photon blockade, analogous to Coulomb blockade in electronic systems, restricts multiple photon occupancy in the cavity, thereby enabling nonclassical light generation such as anti-bunched photon statistics and single-photon sources. On the other hand, the Rydberg polariton state represents a coherent superposition of photonic and collective Rydberg excitations, which under appropriate conditions, can propagate through the medium while retaining strong mutual interaction. These emergent effects form the building blocks for photon-based quantum logic elements, quantum transistors, and non-destructive photon-photon gates, opening the door to scalable and deterministic optical quantum computing architectures.

In this chapter, we will review such optical nonlinearities by starting from the discussion of optical Kerr effects and Rydberg-EIT in cavity-Rydberg system. After this, we will extend our discussion of optical nonlinearities to the Rydberg blockade phenomenon and explain the generation and evaluation of Rydberg polaritons in cavity-Rydberg system.

4.2 Optical Nonlinearities

The concept behind the emergence of optical nonlinearity mediated by Rydberg atoms is based on strong and long-range electric dipole interaction between cavity field and intra-cavity Rydberg ensemble [1]. This interaction is fundamentally governed by the collective excitation of atoms to high-lying Rydberg states, which possess exaggerated electric dipole moments, resulting in pronounced interactions even at micrometer-scale separations. The problem that limit the process is the coherent and localized manipulation of electromagnetic field at the level of dipole-dipole interaction, however, it can be overcome with the principles of cavity quantum electrodynamics (QED) [2]. Cavity QED allows for enhanced light-matter coupling by confining the electromagnetic field within a high-finesse optical cavity, thereby significantly increasing the interaction time and strength between photons and atoms. The cavity QED provides that situation where strong nonlinearity at single-photon range can be obtained by just making system more complicated by associating multiple degree of freedom leading towards the perturbation domain and the broadening of intra-cavity energy spectrum. In such a configuration, the hybridization of photonic and atomic degrees of freedom results in new dressed states and nontrivial quantum correlations that can manifest as nonlinear optical phenomena even at very low light intensities.

Further, the phenomenon of slow light naturally occurs to assist the localized manipulation of light because of the coherent interaction of intra-cavity photons with dense atomic ensemble [3]. The formation of electromagnetically induced transparency (EIT) windows in the presence of Rydberg states significantly reduces the group velocity of the photons, thus allowing for enhanced temporal overlap and controlled interaction between light and matter. The combined state of light and matter, where a dipole moment is excited with the interaction of a single photon, is named as a polariton. These quasiparticles, comprising photonic and atomic excitation, serve as a crucial platform for mediating photon-photon interactions within the cavity. The light-matter interaction ratio along with the dynamics of polariton are defined by the group index n_g of such medium, which should be large in order to coherently manipulate light. A high n_g implies a steeper dispersion relation, thereby enhancing the temporal compression of light and enabling stronger nonlinear interactions within the medium. For the case of Rydberg ensemble, the optical nonlinearities not only depend on the large group index but also on the strong and long-range dipole-dipole interactions, which indeed play a crucial role in cavity QED [1]. These dipole-dipole interactions among Rydberg atoms result in level shifts and blockade effects that are central to the emergence of strong optical nonlinearities.

The optical nonlinearities emerge in matter-wave interaction when the matter interacts differently to the second incident photon as it interacts with the first photon which is possible in two cases. In the first case, the atoms are completely insensitive to the second interacting photon, which means it neither absorbs nor scatters the second photon simultaneously. This arises due to Rydberg blockade, which suppresses further excitations in nearby atoms once one atom is excited to a Rydberg state, effectively making the medium transparent to the second photon. In the second case, the optical nonlinearities rely on the resonant condition for the second interacting photon, which is the fundamental reason behind the optical nonlinearities in cavity-Rydberg system. When the second photon becomes resonant with the shifted energy levels created by the Rydberg excitation of the first photon, a conditional response is generated that leads to strong photon-photon correlations. In cavity-Rydberg mediated optical nonlinearities, the first photon interacts with Rydberg atoms and excites multiple Rydberg states, including the states of neighboring atoms, which lead to the energy shift in whole ensemble. This collective excitation produces a Rydberg interaction volume where subsequent excitations are inhibited. After this, the resonant interaction of second photon causes the excitation of blockaded state. This phenomenon is known as a Rydberg blockade excitation, which in cavity configuration can be extended to the photon blockaded states. This photon blockade effect manifests as anti-bunching of transmitted photons and is a hallmark of strong quantum nonlinearities at the single-photon level. These outcomes in the form of optical nonlinearities are the result of highly sensitive nature of Rydberg atoms to the intra-cavity strongly driven optical field. Such sensitivity not only enhances the nonlinear response but also allows tunability through external control fields, enabling programmable quantum optical devices. In order to develop and enhance our understanding of the cavity-Rydberg optical nonlinearities, we first consider the case of single-photon (or even less than one photon) configuration in the absence and in presence of Rydberg-Rydberg interactions by using classical picture. This simplified model helps build the intuition for more complex quantum mechanical treatments, revealing how even classical analogs of blockade and dipole-dipole shifts can elucidate the physics underpinning emergent nonlinearity in these hybrid systems.

4.2.1 Single-Photon Kerr Effect

The optical nonlinearities emerging from the interaction of the electromagnetic field with atomic ensemble are measured by the energy shift in the response of the atoms. The product of energy shift and the effective dispersion of the output light, which can be determined by the group refractive index of the atomic medium, defines the strength of nonlinearity as,

$$n_g = 1 + \omega \frac{\partial n_i}{\partial \omega}, \tag{4.1}$$

where ω represents angular frequency of incident light and parameter n_i is for the individual refractive index which, for the case of impure medium, can be defined as, $n_i = 1 + \mathcal{X}_r/2$. Here $\mathcal{X}_r$ is the real part of the complex susceptibility of Rydberg medium. The resultant refractive index for the Rydberg ensemble in term of real susceptibility, read as,

$$n_g = \frac{\omega}{2}\frac{\partial \mathcal{X}_r}{\partial \omega}. \tag{4.2}$$

Further, the energy shift, or known as shark shift, in corresponding Rydberg energy spectrum triggered by the incident electromagnetic field depending upon the atomic polarization can be defined as [1],

$$\Delta_{Ryd} = -\frac{\alpha \mathcal{E}^2}{2\hbar}, \tag{4.3}$$

where the parameter α determines the polarization of atomic medium at particular frequency of incident electromagnetic field and is the basic reason behind the scattering and frequency variation of incident field during interaction and possesses key value in generating strong optical nonlinearities. Here it should be noted that, at optical frequencies, Rydberg atoms have very weak polarization characteristics but in the frequency domain ranging from dc to microwave, Rydberg atoms possess very strong abilities for polarization and this characteristics of Rydberg atoms have been beautifully manipulated to ultra-sensitively measure the intra-cavity photon number states [2]. Mostly, the susceptibilities corresponding to the medium with low-frequency response are measured corresponding to the principal quantum number n^7. Further, to include the effects of Rydberg-Rydberg interaction, we consider another nearby Rydberg atom with a similar electromagnetic response $\mathcal{E}$ creating an electric dipole on the scale of principal quantum number n^2. In accordance to these conditions and by using the van der Waals interaction picture, we obtained the dipole-dipole interaction strength on the scale of $\alpha\mathcal{E}^2 \sim n^{11}$ which appears to be in long-range regime. Finally, by manipulating optical response of Rydberg medium and Stark shift produced in atomic energy spectrum, one can obtain the optical nonlinearity which appears to be in the form of optical Kerr effect [1],

$$\mathcal{X}\mathcal{E}^2 = \frac{\partial \mathcal{X}_r}{\partial \omega}\Delta_{Ryd} = -\frac{n_g \alpha}{\omega}\mathcal{E}^2. \tag{4.4}$$

One can categorically note that the strength of optical nonlinearities in term of Kerr effects are directly proportional to the effective group refractive index of Rydberg medium and polarization strength of the Rydberg ensemble. Therefore, in order to obtain large amount of optical nonlinearities, we should have large group index which is only possible in the resonant interaction of intra-cavity field and Rydberg atomic ensemble. The problem that limits the feasibility of large optical nonlinearity in the resonant configuration, is because of the large imaginary part of the susceptibility which causes the off-axis scattering of resonant field. To suppress the imaginary part of the susceptibility, another electromagnetic field is used that generates a transparency on the resonant interaction of Rydberg ensemble and intra-cavity field because of the destructive interference between atomic transition pathways [3]. This phenomenon is interpreted as an EIT and it has made possible to increase effective group index up to the order of 10^6 [4]. Because of such significantly large group index, it is possible to trap electromagnetic pulse of light inside an interacting medium which gives birth to the concept of low-light. This phenomenon of slow-light due to the large group index has contributed a lot in implementation of quantum memory elements by storing and retrieving light pulses [5].

4.2.2 *Electromagnetically Induced Transparency*

As discussed before, EIT arises from the destructive interference between different atomic transition path-ways. Therefore, in order to demonstrate EIT in cavity-Rydberg system, we consider a Rydberg atom inside cavity driven by another probe field which yields in three-level atomic configuration, as illustrated in Fig. 4.1a. To accommodate the enhancement due to large group index, we discuss the derivation of effective susceptibility for intra-cavity three-level system configuration, as discussed in detail in [6] and reviewed in [1, 3]. The external probe laser, with frequency Ω_p, drive the intra-cavity atomic transition from ground state $|g\rangle$ to the first excited level $|e\rangle$ with the detuning Δ_p. The external pump laser or control field, having interaction frequency Ω_c, further excite atoms to the Rydberg state $|r\rangle$ with detuning Δ_c. The time dependent evolution of such three level system by using interaction picture can be interpreted as,

$$\dot{\rho}_{eg} = -i\frac{\Omega_p}{2}(\rho_{gg} - \rho_{ee}) + i\Delta_p\rho_{eg} - i\frac{\Omega_c}{2}\rho_{rg} - \gamma_{eg}\rho_{eg}, \tag{4.5}$$

$$\dot{\rho}_{rg} = i\frac{\Omega_p}{2}\rho_{re} + i(\Delta_p + \Delta_c)\rho_{rg} - i\frac{\Omega_c}{2}\rho_{eg} - \gamma_{rg}\rho_{rg}, \tag{4.6}$$

$$\dot{\rho}_{re} = -i\frac{\Omega_f}{2}(\rho_{ee} - \rho_{rr}) + i\Delta_c\rho_{re} - i\frac{\Omega_p}{2}\rho_{rg} - \gamma_{re}\rho_{re}, \tag{4.7}$$

where terms γ_{eg}, γ_{rg} and γ_{re} correspond to the damping of atomic transition for level $e \leftrightarrow g$, $g \leftrightarrow r$ and $e \leftrightarrow r$, respectively. The damping rate γ_{eg} and γ_{re} are mostly larger than the damping rate γ_{rg} because of the two-photon transition. In the case of high finesse cavity where the only spontaneous emission contribute to the decoherence mechanism than the damping rate $\gamma_{eg} = \Gamma/2$, where Γ accommodates the effects of spontaneous emission corresponding to the first excited state $|e\rangle$. By using mean-field approximation by adopting weak-coupling regime and considering that all atoms are initially in ground state (such as, $\rho_{gg} = 1$, $\rho_{ee} = 0$ and $\rho_{rr} = 0$), we can implement steady-state condition on above differential equations by putting time-derivative equal to zero. It yields in $\rho_{re} = 0$ while using the steady-state solution of

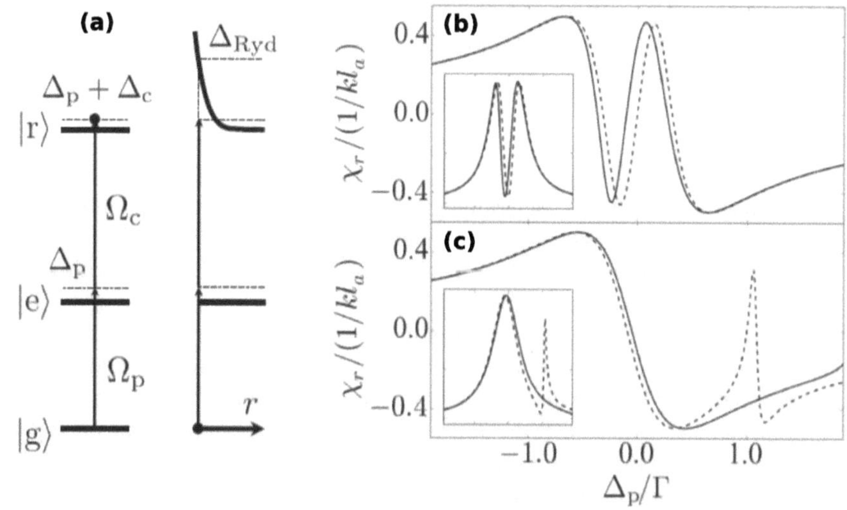

Fig. 4.1 **a** A three-level configuration for the Rydberg atoms, simple (left) and with Rydberg excited blockade (right), excited by the external probe field Ω_p (single-photon excitation) and external pump/control field Ω_c (two-photon excitation) which, due to the Rydberg shifted level, trigger Rydberg blockade excitation. Here Δ_p and Δ_c correspond to the detuning of the probe and control field excitation, respectively, while Δ_{Ryd} is the detuning of the Rydberg sifted levels. **b** and **c** the normalized susceptibilities $\mathcal{X}_r/kl_a$ illustrating the emergence EIT in the dispersion and absorption (inset plots) quadratures of optical transmission field, obtained from the outgoing optical mode. The dashed and solid curves correspond to the responses of three-level setups with simple, for both resonant $\Delta_c = 0$ (**b**) and off-resonant $\Delta_c = -\Gamma_{EIT}$ (**c**) cases, and with shifted Rydberg level, for small shift $\Delta_{Ryd} \sim -0.1\Gamma_{EIT}$ (**b**) and for large shift $\Delta_{Ryd} >> \Gamma_{EIT}$ (**c**), respectively. It can be noted that the EIT-like window disappears in large shift case, case of Rydberg blockade, because of the transformation of the three-level setup into two-level setup. Reused from [1], licensed under CC-BY 4.0 from IOP Publishing

third equation in the steady-state solution of first equation, in order to measure the coherent spectrum in the probe field, we obtained,

$$\rho^s_{eg} = -\frac{i\Omega_p/2}{\gamma_{eg} - i\Delta_p + 4\Omega_c^2/[\gamma_{rg} - i(\Delta_p + \Delta_c)]}. \tag{4.8}$$

From above expression, one can note the emergence of electric dipole moment on the external probe field transition. The resultant susceptibility for such three-level configuration on the probe field reads as,

$$\begin{aligned}\mathcal{X}_{3-level} &= \frac{2\gamma_{eg}\rho_{eg}}{kl_a\Omega_p} \\ &= \mathcal{X}_{2-level} \times \left[1 - \frac{\Omega_c^2}{4(\gamma_{rg} - i\Delta_p - i\Delta_c)(\gamma_{eg} - i\Delta_p) + \Omega_c^2}\right],\end{aligned} \tag{4.9}$$

which appears to be proportional to the bare two-level susceptibility of Rydberg atoms in absence of the probe field frequency ($\Omega_c = 0$), which reads as,

$$\mathcal{X}_{2-level} = \frac{i\gamma_{eg}}{kl_a(\gamma_{eg} - i\Delta_p)}, \tag{4.10}$$

where parameter $l_a = (N\sigma)^{-1}$ determines the distribution length of N atoms resonant to the cavity field and σ is the cross-sectional area for the interaction of intra-cavity optical mode which in the case of two-level configuration is $\sigma = 3\lambda^2/2\pi$ with wavelength λ of the corresponding probe field. These susceptibilities containing the EIT response of the intra-cavity atomic ensemble are shown in Fig. 4.1b and c which describe both absorption (inset plot) and dispersion quadratures of field coming out after interacting with Rydberg ensemble. One can note a dip appearing in both absorption and dispersion due to the destructive interference in atomic transition path from ground state to the excited state and from excited state to the Rydberg state or blockaded state. Such fluctuations in electromagnetic field response of the atomic (or other mechanical objects) driven by an external pump field and a comparatively slow probe field is intuitively referred to the EIT windows.

In order to further enhance our understanding, we discuss EIT behavior by considering the resonant case in which external pump field oscillates at the same atomic transitional frequency, i.e. $\Delta_c = 0$. Here, it should be noted that in the case of two-level system, the scattering of interacting field is maximized, however, in the three-level system, such scattering is suppressed because of emergence of EIT. If we consider the two-photon resonance case, or considering resonance response from both transition $\Delta_p = -\Delta_c = 0$, the susceptibility of three-level configuration will reduce to,

$$\mathcal{X}_{3-level}(\Delta_p = -\Delta_c = 0) = \mathcal{X}_{2-level}\left[1 - \frac{\Omega_c^2}{4\gamma_{re}\gamma_{eg} + \Omega_c^2)}\right]. \tag{4.11}$$

The above expression reveals that the maximum strength of EIT can be achieved if $\Omega_c^2 >> 4\gamma_{re}\gamma_{eg}$, where the maximum length of EIT window can be expressed as, $\Gamma_{EIT} = \Omega_c^2/4\gamma_{eg}$ limited by the broadening condition $\Gamma_{EIT} >> \gamma_{re}$. Intuitively, if we intend to narrow the EIT window to only one-photon or intend to demonstrate a memory element that can store and retrieve a single-photon through EIT then the width of the EIT window should be very less than the damping rate during atomic transition due to probe field ($\Gamma_{EIT} << \gamma_{eg}$). Under these condition, i.e. $\Delta_c = 0$ and $\Gamma_{EIT} << \gamma_{eg}$, the three-level susceptibility will be reformed as [1],

$$\mathcal{X}_{3-level} \approx \mathcal{X}_{2-level}\left[1 - \frac{\Gamma_{EIT}}{\Gamma_{EIT} - i\Delta_p}\right]. \tag{4.12}$$

On the other hand, if we consider resonant case around $\Delta_p = 0$, the real part of the susceptibility will be transformed to $\mathcal{X}_r\Delta_a/kl_a\Gamma_{EIT}$ which reveals the effective group index $n_g \approx \omega/2kl_a\Gamma_{EIT}$. Therefore, in order to obtain large optical nonlinearities to the single-photon regime, it is necessary that the width of EIT window should approach to single-photon which is possible by the manipulation of large group index. To obtain the narrow width of EIT, a smallest value of Γ_{EIT}, it is required that the lifetime of atoms in state $|r\rangle$ should be long, where the state $|r\rangle$ pretend as an analog of the ground state, which is quite possible in the case of Rydberg atoms because of the Rydberg excitation blockade. Such kind of Rydberg EIT was fist experimentally used to demonstrated the Kerr effects due to another external electric field interacting in 2008 [7].

The external probe laser interacts with intra-cavity Rydberg ensemble such that it excite a Rydberg polariton, discussed later in this section, which moves through the EIT window with group velocity $v_g = c/n_g$ which is inversely proportional to the group index. It reveals that for a large value of the group index, the polariton state velocity will be minimum which provides chance to slow down, or even trap, photonic state in a transparency window. As discussed previously, the interaction between Rydberg ensemble and electromagnetic field shifts energy level of Rydberg atoms to certain limit where Rydberg blockaded, as discussed following in this section, state is created which consequently modifies the external pump field detuning with the atomic transition frequency. The shift that trigger blockade excitation are mostly larger than the width of EIT window generating a case where the three-level susceptibility can be transformed to two-susceptibility [8, 9], as illustrated in Fig. 4.1c, where the EIT window has been shifted towards position probe field detuning in both absorption and dispersion. The radius of the blockade sphere, in which the polariton state suppresses the width of EIT, is obtained by the dipole-dipole interaction potential $V(r_b) = 2\hbar\Gamma_{EIT}$ which in van der Walls interactions limits can be defined as, $r_b = \sqrt[6]{c_6/2\hbar\Gamma_{EIT}}$. In a classical regime, or in a semi-blockaded excitation, under the condition of weak probe field $\Omega_c << \Omega_p$, during the transformation of three-level system to the two-level configuration, only a fraction of $(\Omega_p/\Omega_c)^2$ Rydberg atoms in two-level system will get excited which again reforms the three-level susceptibility as [1, 10],

$$\mathcal{X}_{3-level} \approx \frac{4\pi r_b^3 \Omega_p^2}{3\Omega_c^2} N \mathcal{X}_{2-level}, \tag{4.13}$$

where N is the total number of atoms that transform in the two-level setup and external pump field frequency can be redefined as a function of electric field effect of single control photon $\Omega_p = d\mathcal{E}_p/\hbar$, where parameter d represents electric dipole moment for the transition in two-level system. Thus, by analyzing above equation, one can easily extract the Kerr like optical nonlinearity effects. Such classical type of optical nonlinearities was first experimentally demonstrated in 2010 by using $N = 2 \times 10^{10}/cm^3$ Rb Rydberg atoms in a magneto-optical trap [11]. Following, we will further extend our discussion of optical nonlinearities in quantum domain.

4.2.3 Quantum Picture

The ultimate goal in optical nonlinearity is to achieving such regime where the inter-photon interaction are so strong that the number of photons in excited electromagnetic pulse substantially varies with the photon number state [12]. Such optical nonlinearity are only explainable and achievable in quantum domain where the photon-photon interaction have that large value while the intra-cavity photon number states are limited, see Fig. 4.2. Because of the small photon number, quantum picture provides us an opportunity to manipulate strong inter-photon interactions to produce quantum control of electromagnetic field. In this strongly nonlinear quantum regime, even a single photon can significantly alter the optical properties of the medium, which is fundamentally different from the classical nonlinear optics where large intensities are required to induce noticeable changes. The quantum regime offers promising routes toward deterministic quantum gates, quantum memories, and nonclassical light sources.

Although the practical demonstration of quantum nonlinearities is highly difficult because of the less nonlinear response of the materials to the electromagnetic field, however, the realization of quantum nonlinearities is extremely significant in order to enhance classical nonlinear optical phenomena and their applications in communication and computation. Particularly, in cavity or waveguide systems embedded with Rydberg atoms, the large polarizability of high-lying Rydberg states enables enhanced sensitivity to photon-induced atomic level shifts, thereby paving the way for strong nonlinear interactions at the few-photon level. Furthermore, the integration of these systems with photonic platforms is expected to lead to scalable quantum devices that can operate at room temperature with high fidelity.

The interpretation of Rydberg EIT in quantum nonlinearity regime is determine by the strength of detuning between external probe field and atomic transitional frequency Δ_p which indeed cause the broadening of the atomic energy levels and further leads to the quantum interference among different transitional paths. This detuning-dependent behavior is crucial in shaping the transmission spectra, absorption profile, and group velocity of the probe field. As Δ_p increases, the destructive quantum interference becomes dominant, leading to modified transparency windows and altered dispersion characteristics. Such coherent control over light-matter interactions, mediated through EIT and quantum nonlinearities, opens new opportunities in designing quantum optical switches, single-photon transistors, and photonic quantum circuits that rely on Rydberg-mediated long-range interactions.

The interaction of the individual photon with the Rydberg atom excites a Rydberg state to a certain level where the EIT width is suppressed within the excitation level, called as photon blockade sphere or bubble. Here the transmission of other photons through that excited bubble, depending upon the intra-cavity optical depth, can expressed as,

$$T_{2-level}(\Delta_p) = exp\left(- \frac{\gamma_{eg}}{\gamma_{eg} - i\Delta_p} \frac{D_d}{2} \right), \tag{4.14}$$

which is basically equal to the optical response of two-level bare state where the size of the blockaded sphere is determine by the ration of the Rydberg blockade radius r_b and the total attenuation length of the Rydberg ensemble l_a, as $D_d = 2r_b/l_a$. Intuitively speaking, the explanation of quantum nonlinearities of light is obtain from the ratio of the optical response of two- and three-level configuration, computed from classical description of $\mathcal{X}_{2-level}$ and $\mathcal{X}_{3-level}$. In cavity environment, that nonlinear optical response is further characterized in quantum domain by dissipative nonlinearities and dispersive nonlinearities depending upon the resonant and off-resonant configuration, respectively [1].

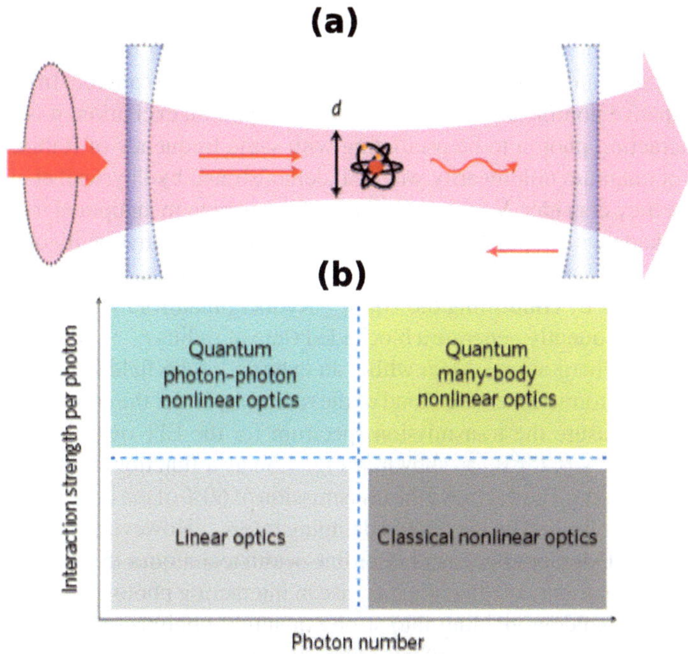

Fig. 4.2 **a** The schematic configuration of a cavity-atom system where the intra-cavity field is confined to the diameter d when it interacts with an atom. In a resonance interaction case, where a single-Photon interacts with an atom with same atomic transitional frequency, the strong optical nonlinearities are generated which consequentially results in strong photon-photon interaction. These optical nonlinearities can be characterized over the interaction strength possessed by the single-photon and the number of interacting photons, as illustrated in (**b**). The few photons having less interaction strength per photon cannot generate any optical nonlinearity and lie in linear regime (light gray box) while a large number of photons but with less interaction strength will result in classical nonlinear optics (dark gray box). Only the small number of photons with large interaction strength per photon can reveal the quantum picture of nonlinear optics (blue box). However, when a large number of photon with large interaction strength will interact with atoms, it will generate a novel regime of strongly correlated many-body systems. Adopted from [12], ©(2025) with permission from Springer Nature

The quantum nonlinearities are characterized on the interaction strength, quantity of interaction between light and matter causing excitation of matter, absorption and scattering, of a single photon by comparing it with a pair of photons (or multiple photons). In a case, where a single photon interacts better than a pair of photon is referred to dissipative quantum nonlinearity. Such type of quantum nonlinearities can be measured by taking scaled two-time second-order auto-correlation of the external probe field after it interacts with atoms

$$g^{(2)}(t_1, t_2) = \frac{\langle n(t_1)n(n_2)\rangle}{[\langle n(t_1)\rangle\langle n(n_2)\rangle]}, \tag{4.15}$$

where $n(t_1)$ and $n(t_2)$ correspond to the photon number at measurement time t_1 and t_2, respectively. The above equation determines that whether the interacting single photon will excite an atom or not or it will just get absorbed or scattered. In order to get better dissipative nonlinearity in the form of single photon excitation, it is necessary for the interacting photon to be oscillating with same frequency of Rabi transition. This type of quantum nonlinearity was well demonstrated by Peyronel et al. in 2012 [13], where they consider $N = 2 \times 10^{10}$ Rb atoms per cm in an optical trap creating photon blockade with depth $D_d \sim 40$ after interacting with intra-cavity optical field, as illustrated in schematic Fig. 4.3a. They chose width of the EIT window around $\Gamma_{EIT} \sim 10\pi$MHz by controlling the $100S_{1/2}$ Rydberg manifold with external pump field which consequently generate a blockaded state of radius $r_b \sim 10\mu$m occupying about 10000 Rydberg atoms. Meanwhile, an external probe field was continuously interacting with atomic ensemble cloud of the radius w_0 under the condition $w_0 < r_b$.

They also measure the transmission spectrum for the EIT over the probe field, which is taken at ~ 0.25pW, as shown in Fig. 4.3b as a function of probe detuning with the cavity mode. They achieve the transmission of 60% of the interacting photons which is basically limited by the photon damping rate γ_{rg}. However, to avoid this limitation, they use Doppler effects and laser line-width techniques in their experiment. From Fig. 4.3b, one can note that the increase in interacting photon rate increases the strength of EIT window yielding into strong quantum nonlinear effects. To further demonstrate quantum nonlinear effects, they computed the correlation $g^{(2)}(t)$ for the photons transmission quadrature as a function of running time and by considering their mutual interactions, as illustrated in Fig. 4.3c. To make the calculation of correlation independent of background noise, they subtract the initial value of correlation $g^{(2)}(0) = 0.04$ from measurements depending upon the background light limitations. These results of transmission correlation verifies, after a short time $t \sim 0.3\mu$s, the transmission quadrature of the EIT for the less interacting Rydberg state $46s_{1/2}$, as shown inset of Fig. 4.3c, in which photon number in transmission follow a saturated behavior which is interestingly appears to be very near to the initial coherent state of the interacting photons. It has been observed that the photon exceeding beyond the photon blockade radius follow anti-bunched mechanism depending upon the length $v_g t \sim 50\mu$s as illustrated by the above axis of the graph which reveals the dissipative nonlinearities beyond the previous concept. They further demonstrated the saturated behavior of the photon leaking out of the cavity and discussed the dependence of

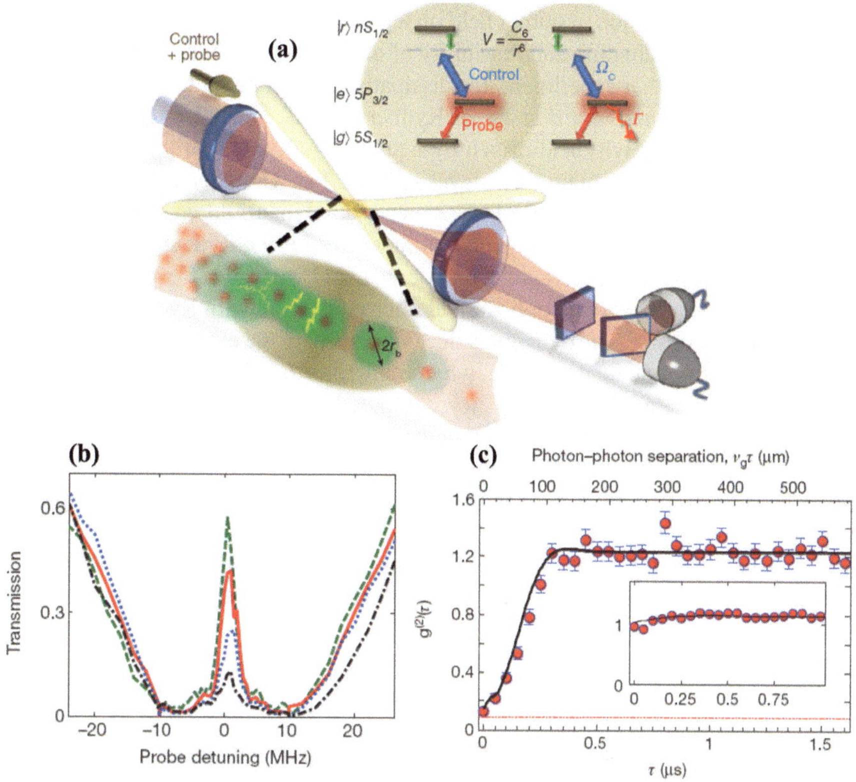

Fig. 4.3 **a** The schematic diagram for the cavity-Rydberg system driven by the external a probe field Ω_p and a control field Ω_c. The inset shaded figure is the level configuration, similar to the configuration shown in Fig. 4.1a, illustrating the atomic transition from state $|g\rangle$ $(5S_{1/2})$ to $|e\rangle$ $(5S_{3/2})$ and from state $|e\rangle$ to Rydberg state $|r\rangle$ $(nS_{1/2})$ with probe and control field, respectively, for $V = C_6/r^6$. The interaction of a probe field with Rydberg atoms in presence of cavity field will excite Rydberg blockade of radius r_b, which afterwards turn into photonic blockade. **b** The transmission spectrum containing EIT as a function of probe field detuning Δ_p obtained from the outgoing optical mode after interacting with the Rydberg state $|100s_{1/2}$. Here dashed green, solid red, dotted blue and dot-dashed block curves correspond to the $R_i = 1, 2, 4$ and 6, respectively, when optical depth is $D_d = 40$ and group delay time is $\tau_d = 250$ns. **c** The photon-photon auto-correlation $g^{(2)}(t_1, t_2)$ of the outgoing optical mode after interacting with the Rydberg state $|100s_{1/2}$. Inset plot shows the auto-correlation for the Rydberg state $|46s_{1/2}$. Here the solid curves correspond to the theoretical calculation which matches exactly with the experimental results. It can be seen that the auto-correlation for the both states show similar saturated behavior. Adopted from [13], ©(2025) with permission from Springer Nature

two-time auto-correlation of output field on the EIT transmission quadrature in order to extend the understanding and feasibility of dissipative quantum nonlinear optics.

Recently, in another experiment, the dissipative type of quantum nonlinear optics has been demonstrated in a d states of Rydberg atoms which are trapped inside a high finesse optical cavity and driven by a single mode external pump field and a probe field, in order to obtain three-level system [14]. They demonstrated that the quantum nonlinearities for the Rydberg states around manifold d are in accordance to the semi-classical picture of optical nonlinearities for the Rydberg states s which the van der Waals interaction regime, in order to avoid blockade effects. To obtain the quantum nonlinearities in the optical field after interacting with Rydberg states d, it is necessary to consider the dynamical decay induced by the excitation blockade due to the intra-cavity optical dark-state, which may referred to the dark-state polariton as discussed later in this section. They computed the transmission spectrum over the external probe field while trigger the transition from Rydberg state $5p_{3/2}$, with manifold $F = 3$, to the state $37d_{5/2}$, with manifold $F = 4$, by manipulating external probe field, as shown in Fig. 4.4a, where EIT transmission spectrum has been plotted against the probe detuning Δ_p. The analytical expression for the effective optical transmission from the cavity, depending upon the single- and two-photon induced transition in Rydberg states, in accordance to the three-level ladder configuration shown in Fig. 4.1a, can be given as [14],

$$T = \left| \frac{\kappa(\Delta_c + i\gamma_{eg} - \frac{\Omega_c^2}{4(\Delta_r + i\gamma_{rg})})}{(\Delta_p + i\gamma_{eg} - \frac{\Omega_c^2}{4(\Delta_r + i\gamma_{rg})})(\Delta_c + i\kappa) - 2\kappa\gamma_{ef}C} \right|, \tag{4.16}$$

where κ is the damping constant or cavity photon line-width and $\Delta_r = \Delta_c + \Delta_p$ is the detuning parameter for Rydberg state defined by the external control field detuning and probe field detuning. The parameters, defined previously, γ_{eg} and γ_{rg} correspond to the line-width or damping rates of the single-photon transition and two-photon transition, respectively. Here parameter $C = g_0^2 N/2\gamma_{eg}\gamma_{rg}$ accounts for the effective coupling of atomic degree of freedom with the cavity mode, where g_0 is usual single-photon Rabi coupling and N is the total number of atoms. They also use above expression to verify their experimental finding, as shown in the Fig. 4.4a. One can observe that, in the absence of atomic ensemble, the cavity transmission spectrum follow Lorentzian distribution and no EIT window appears in the transmission spectrum because of the absence of quantum interference in cavity transition path ways. On the other hand, in the presence of atomic ensemble but the absence of external control field, the cavity transmission spectrum illustrates the intra-cavity optical mode splitting which can be interpreted as the emergence EIT due to the cavity-atom interference, as discussed before in Fig. 4.2.

However, in the presence of external control field that trigger the Rydberg excitation to the Rydberg state $37d$, the cavity transmission quadrature carry another transparency window at two-photon transition, which can be an analog of the double-EIT windows due to the double destructive interference caused by the Rydberg states around $37d$. Such quantum interference can intuitively refer to the Fano interference

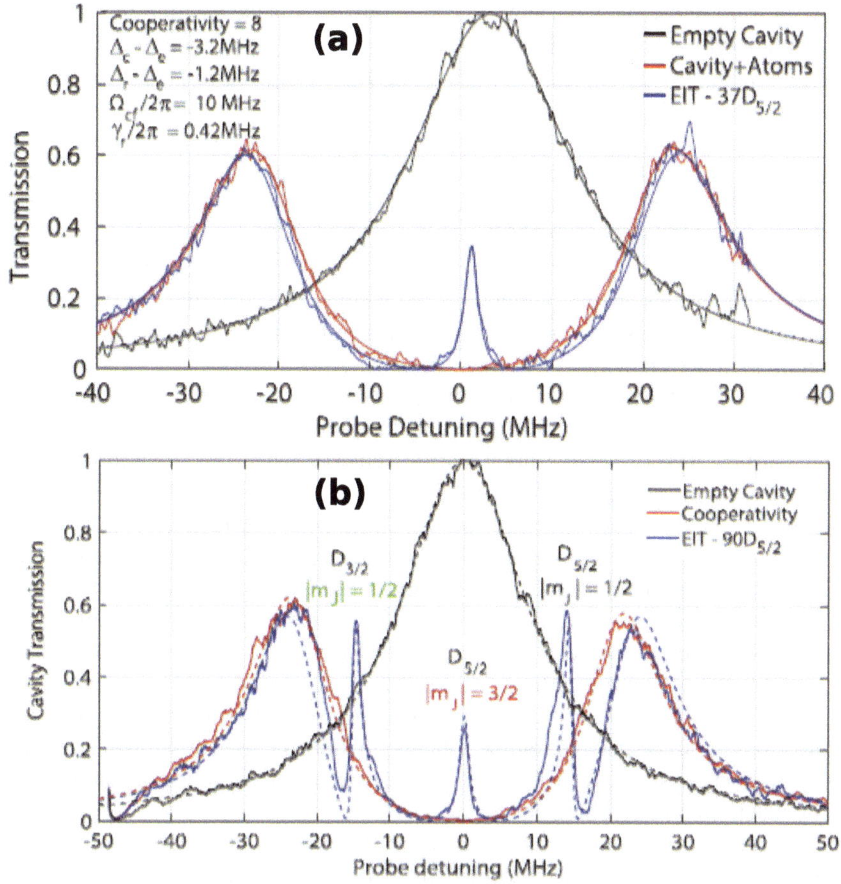

Fig. 4.4 **a** The optical transmission spectrum of the cavity-Rydberg system as a function of probe field detuning Δ_p when the external control field is applied to excite Rydberg state from $5p_{3/2}$ with $F = 3$ to $37d_{5/2}$ with $F = 4$. Here the black curve represents transmission in absence of atomic mode while red curve illustrates the presence of atomic mode and the blue curve corresponds to the emergence EIT due to the Rydberg state $37d_{5/2}$. The experimental configuration provides the choice of multiple parameters as mentioned in the figure. **b** The cavity-Rydberg transmission spectrum when the control field is used to excite the Zeeman sub-levels of the Rydberg state $90d$. Similarly, black curve is for empty cavity while red and blue curves correspond to the co-operatively and EIT quadratures, respectively. The transmission fluctuations correspond to different atomic level excitations denoted with $D_{3/2}$ ($m_j = 1/2$), $D_{5/2}$ ($m_j = 1/2$) and $D_{5/2}$ ($m_j = 3/2$). The interaction of control field with different sub-levels produces multiple splits in the cavity mode. The little mismatch between the empty cavity configuration and atom-cavity configuration is because of the background thermal excitation of atoms. Adopted from [14], ©(2025) with permission from IOP Publishing

between different transitional levels. The double EIT behavior, due to the double quantum interference caused by the interaction of cavity mode with multiple degree of freedom, and its control has been thoroughly investigated in [15]. Further, they demonstrate the cavity transmission spectrum by computing outgoing optical mode when the cavity mode excites multiple Zeeman sub-levels, defined over the magnetic quantum number and electronic manifold, of the Rydberg state $90d$, as illustrated in Fig. 4.4b. It has been shown that the interaction of intra-cavity field with the degenerated Zeeman sub-levels, such as $90d_{3/2}$ with $|m_j| = 1/2, 90d_{5/2}$ with $|m_j| = 3/2$ and $90d_{5/2}$ with $|m_j| = 1/2$, creates multiple splitting of the cavity causing emergence of the multiple transparency window in the output field spectrum. These multiple transparency windows possess key importance in the practical implementation of quantum memory elements.

Another type of quantum nonlinear optics that appear due to the off-resonant interaction of intra-cavity field with trapped Rydberg ensemble is known as dispersive nonlinearities. The dispersive nonlinearities emerge when the strength of single-photon excitation is equal to the excitation induced by the pair of photons. In this case, optical transmission spectrum follow almost same behavior as in dissipative type of nonlinearity but requires a particular optical phase of interaction ϕ different from the previous case and it should be in order of π. Due to the off-resonant probe configuration ($|\Delta_p| >> \gamma_{eg}$) require by the strong dispersive quantum nonlinearity, the bare two-level optical response will yield as a function of that particular phase, $T_{2-level}(|\Delta_p| >> \gamma_{eg}) \approx exp(i\phi)$, where the phase ϕ is defined over the optical depth as, $\phi = -D_d\gamma_{eg}/2\Delta_p$ [1]. One can switch optical nonlinearities from dissipative regime to dispersive regime, obtained from EIT transmission spectrum of cavity-Rydberg system, by just adjusting both external probe field and pump field to excite indeterminate Rydberg states.

The first experimental demonstration of dispersive optical nonlinearity with cavity-Rydberg system was performed by Parigi et al. in 2012 [16], where they place cold Rydberg atomic ensemble in a high finesse optical cavity and measure the dispersive nonlinear behavior by observing the fractional blockade excitations in Rydberg ensemble which lie within the ensemble volume. Soon after this, in 2013, Firstenberg et al. [17] performed another experiment in which they demonstrate quantum dispersive optical nonlinearities in a same cavity-Rydberg system by just choosing the control and probe detuning larger than the damping rate of single-photon excitation $\Delta_p \sim \Delta_c \sim 5\gamma_{eg}$ to obtain the required phase condition. The cavity field transmission obtained from outgoing optical mode (Fig. 4.5a) and the phase shift occurring in the spectrum (Fig. 4.5b) are obtained from the real and imaginary parts of the susceptibility illustrated in Eq. (4.4), respectively. Both the theoretical and experimental results are shown in Fig. 4.5a and b, where the transmission strength and phase spectrum are obtained by the formulas $exp[-kl_a Im(\mathcal{X})D_d]$ and $kl_a Re(\mathcal{X})D_d/2$, respectively. One can observe that the difference between the response of the two-level and three-level configuration only occurs in phase of the outgoing field, which is in the form of shift towards the probe detuning, yielding in a perfect dispersive quantum nonlinear optics scenario.

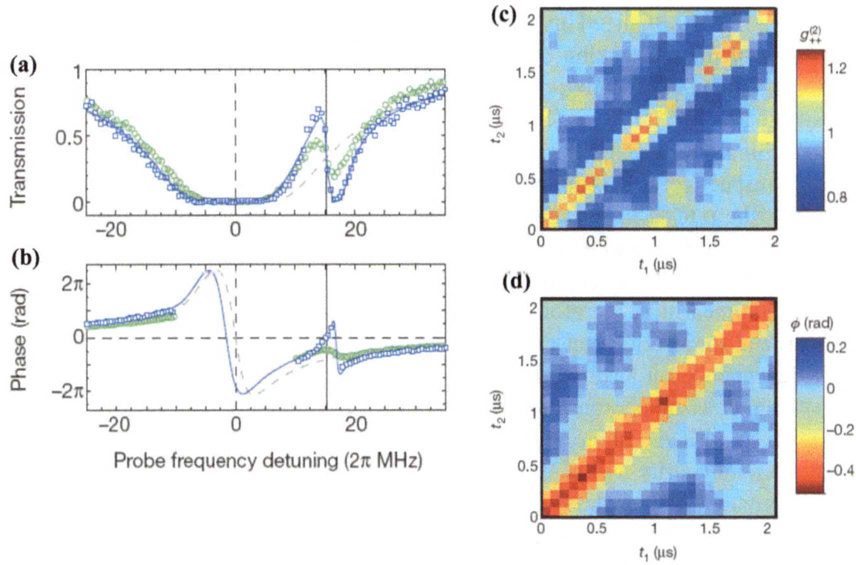

Fig. 4.5 The cavity transmission (**a**) and phase shift (**b**) of outgoing optical mode of cavity-Rydberg system as a function of probe field detuning Δ_p when the control field is detuned to the 15MHz to produce resonance. The blue and green circles correspond to the incoming photon rate $0.5\mu s^{-1}$ and $5\mu s^{-1}$, respectively. The vertical solid black line corresponds to the emergence of dispersive type of quantum nonlinearity. The solid blue and dashed gray curves accommodate the theoretical calculation corresponding to the three-level susceptibility and two-level susceptibility, respectively. **c** and **d** The emergence of dispersive quantum nonlinearity by showing two-photon auto-correlation $g^{(2)}(t_1, t_2)$ measured at time t_1 and t_2. **c** Indicates the photon bunching in the output field while **d** shows the behavior of phase shift of outgoing optical field to reveal the dispersive type of nonlinearity. The color configuration corresponds to the strength of correlation. Adopted from [17], ©(2025) with permission from Springer Nature

To further extend the understanding of quantum dispersive nonlinearities, it's necessary to measure the effects of second photon excitation due to the external control which basically trigger the Rydberg blockade excitation. As discussed previously, these influences can be observed by taking two-photon auto-correlation of the cavity transmission spectrum. As, for the dispersive type of quantum nonlinearities, it is necessary for the outgoing optical mode to fulfill the phase condition rather than transmission spectrum, therefore, it's necessary to implement interferometric mechanism [1] which is possible with the inclusion of another optical mode as a reference with either different polarization or different frequency such that it yields in linear transmission spectrum. One can obtain the density matrix $\rho(t_1, t_2)$, corresponding to the two-photon measurement at time t_1 and t_2, by manipulating two-photon auto-correlation function with the reference optical mode by utilizing quantum tomography method [17]. To interpret an interaction picture between single-photon and two-photon transmission, one can take the tensor product of the density matrices $\rho(t_1) \otimes \rho(t_2)$ corresponding to both transmissions which reveals in a normalized interaction matrix $\tilde{\rho}(t_1, t_2)$ equivalent to the correlation $g^{(2)}(t_1, t_2)$, discussed previously. The normalized interaction matrix not only reveals the information about the phase of photon leaking out of the cavity but also demonstrates the decoherence and entangled dynamic of the system. The behavior of interaction picture of density matrix $g^{(2)}(t_1, t_2)$ is illustrated in Fig. 4.5c and d, where the dynamics of the bunching photons and the implementation of phase-shift condition are demonstrated, respectively. It is found that the phase-shift in the output field spectrum occurs at the rate of $|\phi| = \pi/4$ in reference to the 50% linear transmission which reveals the emergence of dispersive nonlinearity [17]. Further, as it is necessary for the Rydberg-Rydberg interaction to be in strong regime in order to achieve strong optical nonlinearities, therefore, the quantity of photon-photon interaction is also very crucial to describe and characterize nonlinearities in quantum picture due to their dependence on Rydberg-Rydberg interaction. For instance, if the interaction between interacting photons are dissipative then the quantum nonlinearities will be dissipative and if the photon-photon interaction is dispersive, likewise the type of quantum nonlinearity will be dispersive.

Further, they demonstrate the dynamic of dispersive quantum nonlinearity by accommodating photon-photon interactions [18, 19] and their response to the multiple atomic and photonic correlated states [20, 21]. These quantum nonlinearities have been recently studied under the influence of Rydberg-Rydberg interaction and by following perturbation approach to excite the intermediate Rydberg state to produce strong nonlinear effects [22]. Furthermore, the concept of single-photon induced quantum nonlinearities was extended to the creation and implementation of photon blockades [23], which basically correlates with the single-photon excitation of polariton state. The realization and implementation such photon blockade possesses key role in demonstration of quantum information and communication and, therefore, such photon blockade has been extensively studied over the few years [24–26]. The photon blockade excitation in cavity-Rydberg system has been investigated by manipulating the Laplace transform of two-photon relative wave function to obtain the correlation for single- and two-photon resonances [27, 28] and by manipulating EIT to perform optical switching [29]. The quantum nonlinearity, produced due to the excitation of photon blockade, have been discussed recently by manipulating EIT in strong coupling regime which triggers the excitation of dark-state polaritons, as discussed in remaining of this section [30].

4.3 Cavity-Rydberg Blockade

The reason behind the key importance of the Rydberg atoms in coherent implementation of quantum information concepts and in many other phenomena of modern physics lie behind the blockade characteristics which significantly enhance the optical nonlinearities, as discussed previously. When an ensemble of Rydberg atoms interact with cavity mode, driven by a strong external field, resonant to the transitional frequency between Rydberg states, it will trigger excitation process initiating from single atom to all atoms in the ensemble. At this stage, if the inter-atomic Rydberg interactions energies are strong enough, multiple excitations of the Rydberg states will be suppressed and consequently only one atom can get excited. Such blocked excitation of multiple Rydberg atoms corresponds to the Rydberg blockade regime.

If we consider an ensemble of Rydberg atoms with averaged atomic number $N = 3$, then the resultant two-level system in strong interaction regime can be modeled with the ground state $|ggg\rangle$ and with the effective excited state can be defined as $|\mathbf{S}\rangle = (|gge\rangle + |geg\rangle + |egg\rangle)/\sqrt{3}$. Here the coupling between the intra-cavity optical mode and the atomic ensemble is appeared to be in order of $\sqrt{N}$ which is much larger than the intra-cavity field coupling with single atom or single photon Rabi coupling. As these excited states of Rydberg ensemble are get shifted because of having resonant transition with intra-cavity field forming a collective blockaded state in which whole system effectively behaves like a two-level system, therefore, it can explained and evaluated with the help of Jaynes-Cummings model (JCM). Further because of collective enhanced factor $\sqrt{N}$, the strong coupling regime between Rydberg atoms and cavity mode can be achieved which appeals a great interests to the wide variety of quantum mechanical applications. Recent experimental demonstration of Rydberg blockades in strong coupling regime add further to the importance of these phenomena [31–33].

The explanation of the Rydberg excitation blockade in cavity-Rydberg system with the help of JCM was demonstrated by Guerlin et al. [28], in strong coupling regime because of the coherently enhanced coupling between Rydberg ensemble and intra-cavity field, with strength larger than the cavity decay rate. They proposed that in absence of remote Rydberg-Rydberg interaction, the cavity-Rydberg interaction can generate long-range excitation blockade that may have range even larger than the dipole-dipole interaction length. The dipole-dipole interaction effects shifts the energy of the excited state over the size of the Rydberg ensemble which is in order of the length l_R where the strength of dipole-dipole interaction is equal to intra-cavity atom-field interaction. Therefore, during transition $|g\rangle \rightarrow |e\rangle$, it is possible to couple collective ground state of all atoms in ensemble $|G\rangle = |g_1, ..., g_N\rangle$ to the blockaded excited state $|E\rangle = \hat{S}^\dagger|G\rangle$ resonant to the intra-cavity optical mode. The JCM provides an important tool to interpret the dynamics of such effective two-level systems, in the form of a blockade (or a superatom) [34], resonantly interacting with cavity mode. Therefore, the effective Jaynes-Cummings Hamiltonian for cavity-Rydberg blockade under rotating frame approximation and unitary transformation with $exp[-it\omega_c(\hat{a}^\dagger\hat{a} + |E\rangle\langle E|)]$ can expressed as,

$$\hat{H}_{JC} = \hbar\omega_c\hat{a}^\dagger\hat{a} + \hbar\omega_c|E\rangle\langle E| + [\hbar g_{eff}\hat{a}|E\rangle\langle G| + H.c.], \tag{4.17}$$

where $\hat{a}^\dagger$ ($\hat{a}$) are the intra-cavity photon creation (annihilation) operator with cavity mode frequency ω_c while $g_{eff} = g_0\omega_a\sqrt{N}/\Delta$ is the effective coupling of Rydberg ensemble with intra-cavity field defined by the collective enhanced factor $\sqrt{N}$. The Rydberg-Rydberg interaction shift induced in the ensemble is appeared to be in order of $\Delta_{R1} \sim 200 \times 2\pi$MHz which indeed is greater than the strength of effective coupling g_{eff} revealing the Rydberg blockaded state with two or more excited atoms having radius $l_r \sim 7\mu m$ larger than the ensemble size $\sim 4\mu m$.

The coupling between Rydberg ensemble and intra-cavity field can be further manipulated to extend and enhance Rydberg blockade mechanism even beyond the limit of dipole-dipole interaction length. In situation where inter-atomic distance is larger the sample length l_R which suppresses the inter-atomic interaction, the both atomic degree of freedom and cavity mode will behave as two coupled oscillators generating two intra-cavity eigenmodes with frequency $\omega_c \pm g_{eff}$, as illustrated in Fig. 4.6a. If we drive cavity-Rydberg system with another external probe classical field resonant to one of these frequencies, a coherent effective optical mode can be excited which have an eigenmode as a combination of all intra-cavity coherent optical modes and, on the other side, the effective atomic mode accommodates all the excited Rydberg states, forming a atomic bubble. Inversely, if the dimensions of the Rydberg ensemble are equivalent to the l_R, the Rydberg-Rydberg interaction will become so strong that they may shift a friction of Rydberg excited states far from the cavity resonance. Such two-level Rydberg blockade along with the entire cavity-Rydberg system can be explained by the JC Hamiltonian Eq. (4.17), see Fig. 4.6b. Further, if we consider a case where the atomic ensemble size is greater than l_R but dipole-dipole interaction still contribute significantly, then the strength of interaction cannot shift all of the multiple excited Rydberg states our to the cavity resonance which leads to obtain unfinished doubly excited Rydberg states, read as,

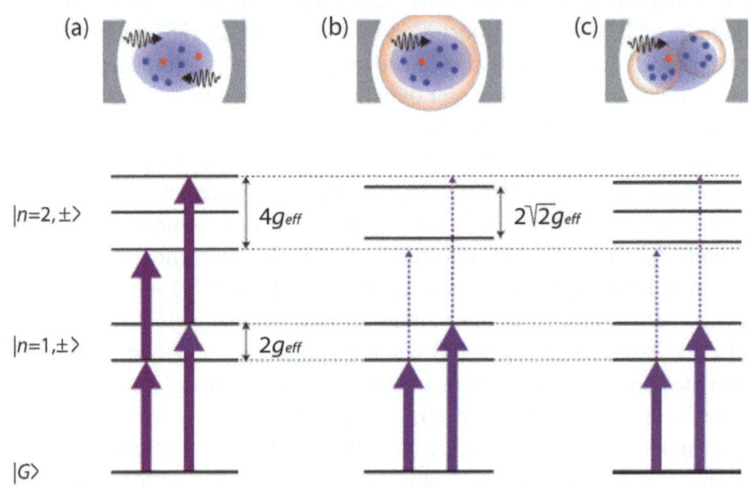

Fig. 4.6 Ladder configuration for the blockade excitation in a cavity-Rydberg system. **a** The transitional configuration in absence of the Rydberg-Rydberg interaction between states $|g\rangle \rightarrow |n=1, \pm\rangle$ (with detuning $2g_{eff}$) and $|n=2, \pm\rangle \rightarrow |n=2, \pm\rangle$ (with detuning $4g_{eff}$), where intra-cavity optical field and atomic degree of freedom behave as two coupled oscillators and corresponding excitation frequencies are degenerated. **b** The influence of dipole-dipole interactions and corresponding level configuration showing blockade formation with shift $2\sqrt{2g_{eff}}$. The dipole-dipole interactions of Rydberg ensemble are so strong that the whole ensemble is fully blockaded and further simultaneous excitation of multiple atoms are stopped by the transitional nonlinearities. **c** The intermediate case, where the size atomic ensemble is considerably large than the blockade size because of weak dipole-dipole interactions exciting multiple blockaded bubbles in which the whole system is interpreted as a combination of dress states. The formation of two bubbles is illustrated having $\sqrt{2g_{eff}}$ shift for each. Adopted from [28], ©(2025) with permission from American Physical Society

$$|E_2\rangle = \frac{1}{\sqrt{\mathcal{A}}}\Bigg(\sum_{|\mathbf{r}_j-\mathbf{r}_k|>l_R} |e_j, e_k\rangle\langle g_j, g_k|\Bigg), \tag{4.18}$$

where the atoms in j, k are separated such that they cannot possess strong mutual Rydberg-Rydberg interaction. The parameter $\mathcal{A}$ is a normalization constant that accommodates number of excited states, a pair of Rydberg atoms whose inter-atomic distance is larger than the interaction length l_R, that contribute to the collective blockaded state $|E_2\rangle$. This state is resonantly connected to the first excited collective state of Rydberg atoms $|E\rangle$ through the intra-cavity initial two-photon absorption process and via external blue detune laser field.

Therefore, the JC Hamiltonian Eq. (4.17) must further accommodate following terms in order to induce second collectively excited state $|E_2\rangle$ [28],

$$2\hbar\omega_c|E_2\rangle\langle E_2| + \hbar g_{eff}\hat{a}\langle E_2|\hat{S}^\dagger|E\rangle\langle E| + H.c.. \tag{4.19}$$

Here one should note that because of the Rydberg ensemble spatial size inside cavity, the resonant interaction between intra-cavity optical field and atomic ensemble in presence of external probe field may excited higher order blockaded states ($|E_3\rangle$, $|E_4\rangle$, ...) corresponding to the number of excited Rydberg states ($n_e = 3, 4, ...$) having spatial distribution larger than the dipole-dipole interaction length l_R. Thus, this process may lead to the decomposition of Rydberg blockade to the number of small blockaded states or bubbles having radius in order of l_R [35]. In such bubbling mechanism, each bobble contains only a singly excited Rydberg state, therefore, the number of atoms excited in Rydberg ensemble is proportional to the number of excited bubbles $n_b \sim 3V/4\pi l + R^3$. In order to accommodate all bubble transition or sub-Rydberg blockades, the system Hamiltonian Eq. (4.17) must contain term $(n_b - 1)$ which determines the coupling between blockaded excited states (i.e. the coupling between $|G\rangle$ and $|E\rangle = |E_1\rangle$, $|E_1\rangle$ and $|E_2\rangle$,... $|E_{n_b-1}\rangle$ and $|E_{n_b}\rangle$).

The multiple Rydberg bubbles in fact is a crude definition of the multiply excited Rydberg states due to the continues nature of Rydberg-Rydberg interactions. However, the initial transitional pair containing the excitation from ground state $|G\rangle$ to the first excited blockade $|E_1\rangle$ constitute as a two-level system due to the direct interaction with intra-cavity field and the higher order excitation can be transformed to the higher-dimensional subspaces having states description $|E_{k>1}\rangle$. By dressing atomic ensemble with the external driving laser, which must be resonant with the initial excited state $(|G, 1\rangle \pm |E_1, 0\rangle)/\sqrt{2}$, it is possible to suppress the higher order excitations. Here, if the second excitation blockaded manifold contains a state that have transitional frequency double than the incoming photons than the photon immediate after the first photon of driving field will be absorbed in order to limit the excitation. It is only possible if the inter-atomic distance is very large so that they do not interact with each other which, consequently, decreases the coupling between the atoms in first excitation blockade and may allow doubly excited state with very different values because of the two-atom excitation.

To get a better estimate of intermediate mechanism that limits the blockade excitation to the first order, the effective cavity-Rydberg system for blockaded states is expressed as,

$$\hat{H}_{JC} = \hbar\omega_c\hat{a}^\dagger\hat{a} + \hbar\omega_c\left(\sum_{k=1}^{n_b} k|E_k\rangle\langle E_k|\right)$$
$$+\hat{a}\left(\sum_{k=1}^{n_b} g_k|E_k\rangle\langle E_{k-1}|\right) + H.c., \tag{4.20}$$

where g_k defines the coupling of each bubble with the intra-cavity field by interpreting each bubble as a two-level system having unique blockade state $|E_k\rangle$ precisely differentiated by the parameter k. These states possess modified the angular momentum state $|j, m\rangle$ which is defined in term of effective modified quantum numbers $j = n_b/2$ and $m = -n_b/2 + k$. The coupling g_k can be defined by manipulating angular momentum creation operator with the effective intrinsic coupling of each bubble $\langle k|\hat{j}^\dagger|k\rangle$ as [28],

$$g_k = \sqrt{(n_b - k + 1)k}\sqrt{\frac{N}{n_b}\frac{g_0\omega_a}{\Delta}}, \tag{4.21}$$

where, as usual, g_0 is the single photon Rabi frequency and ω_a is the intrinsic transition frequency of each bubble. For $k = 1$, Eq. (4.21) reveals the effective coupling g_{eff} for the first excitation blockade and if we set $k = 2$, it shows the coupling for the second blockade, where if we chose $n_b = 1$, it will be in order of $\sqrt{2}g_{eff}$ interpreting unblocked excitation limit over n_b [28]. It means that the value of n_b is not too large, the concept of two-coupled oscillator cannot be implemented and, consequentially, equidistant ladder configuration for the excitation spectrum cannot be used.

The transition during the first excitation is considered red detuned to the transition towards the next excitation step which transforms Rydberg excitation blockade into different bubbles spatially distributed over the cavity length, see Fig. 4.6c. If the excited pairs in Rydberg ensemble follow Förster interaction picture in which they have positive interaction energies, the intra-cavity field will also be red detune due to the interaction with excited pair which will cause the excitation of only one initial dress state resonant to the cavity field. Here it should be noted that the cause behind the absence of further excitation bubbles is not their inter-atomic interactions, it is because of the off-resonant coupling between higher excited states with cavity field emerging from the number of Rydberg pairs in the excited state. The eigenvalues for such blockade excitation can be obtained by the diagonalization of the Hamiltonian 4.20 which read, $0, \pm\sqrt{2 - 1/n_b}\hbar\sqrt{2}g_{eff}$ which illustrate that the intra-cavity field, behaving resonantly with the first dress state $|1, -\rangle$, is red detuned by the amount equal to the two-photon resonance $\sim g_{eff}/2n_b$.

Further, single-photon switching based on Rydberg blockade in cavity environment has been experimentally investigated [29]. The external driving electromagnetic pulse are stored in intra-cavity Rydberg excitation blockade, with the help of electromagnetically induced transparency, which facilitates in transmission of desired electromagnetic pulse generating an optical switch that emits and store particular targeted photon. Because of the strong and long-range dipole-dipole interaction between Rydberg ensemble, the excitation of Rydberg blockade constitute a significant role in understanding quantum state synthesis of the atoms in cavity quantum electrodynamics [36]. It has been numerically shown that the non-classical state of light or photon number state can be obtained because of the coherent interaction between intra cavity optical field and Rydberg excited blockade. The excitation of two Rydberg states within the blockade volume has also been investigated. Furthermore, recently, an inelastic photon scattering in an optical cavity with Rydberg blockade has been demonstrated [37] by manipulating Rydberg electromagnetically induced transparency, which will be discussed in detail in following section. They illustrate the elastic as well as inelastic scattering of intra-cavity optical field from the highly excited Rydberg blockaded state by obtaining the optical field leaking out of the cavity which helps in interpreting Rydberg polaritonic structures, discussed as following. Thus, these cavity mediated Rydberg excitation blockade possesses a key importance in interpreting controlled quantum optics and quantum information [38] and will stand in the basis of majority of the topics that will be discussed in the remaining of this book.

4.4 Rydberg Polaritons

Polaritons are the type of quasiparticle that emerge from the strong and coherent interaction of electromagnetic field with atom(s) generating an electromagnetic dipole moment, which basically triggers the excitation of polariton quasiparticle [33, 39, 40]. These quasiparticles, due to having combine state of light and matter, have a significant importance in modern physics, specifically in strongly correlated many-body system because they have the ability to exhibit the properties of interaction systems like other quasiparticle. The polariton states of excited atoms with light provides an opportunity to store and retrieve photon leading towards quantum memory elements and their inherent photonic structure allow us to use them for information transformation at long distance which enhance the optical quantum communication. Depending upon the interactions of light and boundary configuration, these quasiparticles can be further characterized into multiple types. In the case of semi-conducting micro-cavities, the coherent mean-field interaction of cavity photon [41] number with the cavity-exciton component generates exciton-polariton [42, 43], whose mass and intra-cavity optical evolution arises from the optical degree of freedom. Another type is magnon polariton emerging from the electromagnetic hyperfine excitation of atomic ensemble and are extremely important for long-living storage devices in quantum information [44, 45]. Further, the excitation of surface plasmons interacting with light generates the surface-plasmon-polariton [46], where the optical component provides the manipulation of plasmons which, thus, results in strongly interacting hybrid single photon emitter.

The emergence of highly excited Rydberg state due to the resonant interaction with monochromatic light forms a Rydberg polariton [11, 47] which possess a direct significance in implementation of quantum nonlinear optics [12, 13, 17]. The Rydberg polaritons are extremely useful for the implementation of single-photon devices [29, 48, 49] for quantum computation due to the strongly interacting and highly excited nature of Rydberg atoms while optical characteristics of Rydberg polaritons allow us to use them as a carries in long-distance quantum communication channels [50]. In regards of quantum nonlinear optics, Rydberg polariton play a vital rule in realization of many nonlinear phenomena like EIT, Rydberg blockade excitations and photon-photon interactions. Rydberg polariton and their characteristics have been extensively studied in many system configurations including both free-space and trapped configuration, where the optical component of polariton has been manipulated to control the polariton-polariton interactions which are crucially important to obtain dispersive type of quantum nonlinear optics [14, 16], as discussed previously. Further, these polariton excitation have been studied to manipulate photonic effects in Bose-Einstein condensate (BEC) [51], photonic crystallines [52] and even in topological states emerging form the synthetic magnetic fields [53–56]. In nonlinear optics, when the control field interacts with Rydberg atoms, it excites Rydberg polariton having both optical and Rydberg characteristics. After this, when the control field interacts with the atoms, Rydberg polariton will reduce the speed of probe field, as discussed previously, and excite EIT corresponding to the dark-state of the polariton. Recently, these Rydberg polariton are being the subject of many interesting investigation including the three-body interaction effects on the slow-light dynamics in Rydberg polariton [57], explanation of Rydberg polariton under mean-field theory [58] and the excitation of Rydberg polariton under thermal vapor configuration [59].

The cavity quantum electrodynamics provides most promising approach to enhance the atom-photon interactions and it has not only been proven as a perfect tool to characterize and manipulate highly excited Rydberg atoms but leads us towards the single-photon nonlinear quantum optics. The exploitation of polariton quasi-particles in cavity environment has been very crucial to understand and implement many phenomena of quantum mechanics and quantum optics. In this context, cavity induced nonlinear quantum optics have been demonstrated, even at single-photon level, by manipulating Rydberg polariton [13]. Cavity polariton can be characterized into two types depending upon the transmission of interacting light. If we consider the cavity-Rydberg system driven by a control/pump field and a probe field which lead to the three-level ladder configuration as shown in Fig. 4.1a, then the effective real part of the eigenvalues, corresponding that particular cavity-atom system, can be expressed as, $\lambda = 0, \sqrt{g_0^2 N + \Omega^2/4}$, where as usual g_0 is the single photon Rabi coupling while the Ω is the effective Rabi transitional frequency of the N atoms [60].

The cavity optical transmission carries three relative optical modes corresponding to these eigenvalues, as shown in Fig. 4.7a, where one can observe tree peaks as a function of probe field detuning. The central peak $|D\rangle$ corresponding to $\lambda = 0$,

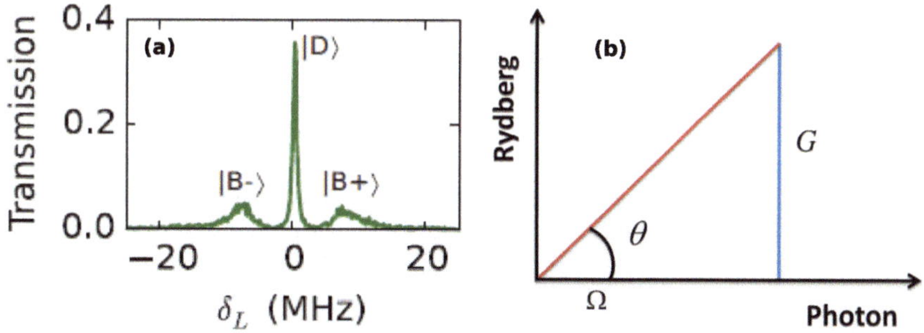

Fig. 4.7 **a** Cavity-Rydberg transmission spectrum as a function of intra-cavity detuning δ_l obtained from the outgoing optical mode of fractional empty cavity. The peak $|D\rangle$ corresponds to the intra-cavity dark-state polariton while peaks $| \pm B\rangle$ represent two cavity bright polariton states. **b** Dynamical decomposition of cavity-Rydberg dark-state polariton into intra-cavity photon and Rydberg excitation component, where angle θ is defined by effective atom-field coupling G and collective atomic transitional frequency Ω. From [69], ©(2025) with permission from American Physical Society

appearing at zero probe field detuning, represents the dark-state polariton which basically emerges because of the EIT. The symmetric side modes $|B\pm\rangle$, corresponding to the non-zero eigenvalues at $\Omega \to 0$, represents the emergence of two bright-state polaritons which appear broader than the EIT peak over the probe detuning spectrum. As these bright-state polariton are unable to fulfill the Doppler-free condition, therefore, they cannot fully stop the incoming photon which is fundamentally required by the EIT. These both bright- and dark-state polariton have been extensively studied in cavity environment and are the subject of many recent investigations. As the dark-state polaritons, due to their direct application in quantum nonlinear optics, possess a significant importance in modern quantum applications, therefore, following we will discuss in detail the cavity-Rydberg dark-state polaritons [61–63].

4.4.1 Dark-State Polaritons

The excitation of intra-cavity dark-state polariton, due to the resonant interaction of cavity field and Rydberg ensemble, develop a scenario where a coherent manipulation of quantum can be perform because of the long lifetime photon from dark-state polariton, which are highly controllable with external control field, and because of the blockade nature of Rydberg atoms which allow us to model whole ensemble as a two-level system. Along this avenue, thus, the complete system can be interpreted by JCM, as discussed in detail previously, and the strong coupling regime can be achieved depending upon the critical amount of photon n_0 and critical number of atoms N_0. These feature allow us to demonstrate and implement state of the art cavity quantum electrodynamics [2]. When quantum state of light that take the form of dark-state polariton during interacting with Rydberg atoms, its group velocity can be decreased because of the control field excitation which results in EIT. This feature was first demonstrated in 2000, by Fleischhauer and Lukin [64] and has been to develop photonic memory elements [65], optical filters at single-photon level [66], and optical switches [67]. The observation and characterization of dark-state polariton in cavity-Rydberg system have been recently investigated theoretically by Lin et al. [68] and experimentally by J. Ningyuan et al. [69]. They both have investigated the cavity-Rydberg excitation and evolution of dark-state polaritons by measuring the optical transmission from cavity.

By following these well-established theoretical as well as experimental investigations, the effective coupling between intra-cavity dark-state polariton, excited by single-photon, and excited Rydberg blockade, which in-fact appears as a two-level system, can be obtained by following JCM under appropriate condition.

$$\hat{H}_{JC-int} = g_0(|\uparrow\rangle\langle\downarrow|\hat{b}_0 + |\downarrow\rangle\langle\uparrow|\hat{b}_0^\dagger), \tag{4.22}$$

where g_0 defines the strength of coupling between optical mode and atomic degree of freedom depending upon the single-photon Rabi transition which most larger than the cavity decay rate and the effects of spontaneous emission in order to achieve strong coupling regime. The operator $\hat{b}_0^\dagger$ ($\hat{b}_0$) corresponds to the creation (annihilation) operators of the dark-state polariton which is responsible for the excitation (absorption) of the intra-cavity polariton. Further, $|\uparrow\rangle\langle\downarrow|$ and $|\uparrow\rangle\langle\downarrow|$ represent flip operator for effective two-level system, formed by the Rydberg blockade excitation, which indicate blockade raising from lower state to the excited state and decay from excited state to the lower state, respectively. The direct advantage of such cavity-Rydberg system lies behind the long lifetime manipulation of intra-cavity photon which basically results from the decrease in the group velocity of back and forth bouncing photons inside cavity because of EIT. Consequently, it will cause the decrease in the line-width of the dark-state polariton resonating inside cavity which makes possible to observe quantum nonlinearities at single-photon level. It is possible by the controlling quantum Rabi transition in Jaynes-Cummings ladder configuration with help of control field which intuitively results in highly non-classical photonic states.

If we consider the effective state of the cavity photon as $|n\rangle$ and the Rydberg blockade state, excited by the same cavity photon, as $|r_c\rangle$ defined by the collective coupling G, then the effective state of cavity-Rydberg dark-state polariton can be expressed as, $|D\rangle = cos(\theta)|n\rangle - sin(\theta)|r_c\rangle$. Here θ is the rotation angle of the intra-cavity dark-state plariton defined by the ratio of collective coupling of Rydberg blockade with cavity mode G and the effective transition frequency of the atomic mode Ω (see Fig. 4.7b), as $\theta = tan^{-1}(G/\Omega)$, where collective blockade coupling can be written under condition $N >> 1$ as $G = \sqrt{\sum_{j=1}^{N} g_j^2/N}$. The angle θ basically is responsible for the quantum phase evolution of the cavity-Rydberg dark-state polariton which leads towards description of quantum phase transition, will be discussed later in this section. Further, one can construct a superposition state of cavity-Rydberg state, also known as Schrödinger-cat state, by simply synthesizing JCM. If we consider the initial state of atomic mode as a superposition state $|g\rangle + |r_c\rangle/\sqrt{2}$ which is a combination of ground state and blockaded state of atoms having equal probability amplitudes and the initial cavity field in a vacuum state $|n_0\rangle$, then the classical probe field, exciting dark-state polariton, can be described by the Hamiltonian $H_p = E^*\hat{b}_0 + E\hat{b}_0^\dagger$, where E corresponds to the intensity of the probe field. Here, one should be remembered that for initial atomic state $|g\rangle$, under condition for the single-photon Rabi coupling $g_0 >> \kappa_0, \gamma$, external probe field cannot be able to excite dart-state polariton because of the normal mode splitting.

However, for the atomic blockade state, Jaynes-Cummings interaction will eventually excite dart-state polariton but under condition $|E| >> \kappa_0$, where κ_0 is vacuum state cavity decay. In this dark-state polariton will consequently appear as a coherent state $|\alpha\rangle$, where $\alpha = Et$ with time t as a probe field interaction time. In accordance to the blockade condition and polariton condition, the final Schrӧdinger-cat state for the cavity-Rydberg state can be written as [68],

$$|cat\rangle = \frac{|g\rangle|n_0\rangle + |r_c\rangle|\alpha\rangle}{\sqrt{2}}. \tag{4.23}$$

Such kind of atom-field superposition states are crucially important not only to understand the basic concept of non-locality in quantum physics but also to the implementation of modern quantum communication.

In experiment [69], they measure the cavity transmission spectrum by exciting the Rydberg atomic state from $|g\rangle = 5s_{1/2}$ with manifold $F = 2$ to $|e\rangle 5p_{3/2}$ with manifold $F = 3$, with the help of external pump field, which eventually results in principal quantum number 40. The measured cavity-Rydberg transmission spectrum is shown in Fig. 4.8a–d, as function of external probe field detuning, where they demonstrate the dynamics of dark-state polariton by observing EIT. By consider low principal quantum number and low intra-cavity photon density, they have construct a regime where the polariton-polariton interaction are weak enough to be ignored in order to observe dynamics of individual polaritons. By observing EIT in cavity transmission spectrum, they measure the energy of dark-state polariton which appears in the order of $\hbar\Delta_D$, where Δ_D is the polariton detuning corresponding to the Rydberg mode, and determine the decay rate of polariton γ_D to calculate their intra-cavity lifetime. Figure 4.8a–d illustrate the cavity-Rydberg transmission spectrum for the different dark-state polariton rotation angles. One can observe that the dark-state polaritons get further slowdown as compare to the previous cavity-polariton observation [62] because they spend more time in exciting Rydberg blockade. It is observed that the energy of the dark-state polariton transforms more slowly as compare to the cavity decay rate depending upon the rotational angle of the dark-state polariton, $\hbar\Delta_D \approx \hbar[\Delta_c cos^2(\theta) + \Delta_{ryd} sin^2(\theta)]$, as shown in Fig. 4.8d.

In order to achieve quantum nonlinearities at single-photon level, it is necessary for photon-photon interaction energy most be larger than the width of intra-cavity dark-state. Therefore, lifetime of polariton possess significant importance not only in regards of obtaining stable optical materials to perform strongly correlated many body physics but also in order to enhance quantum nonlinear optics. In this avenue,

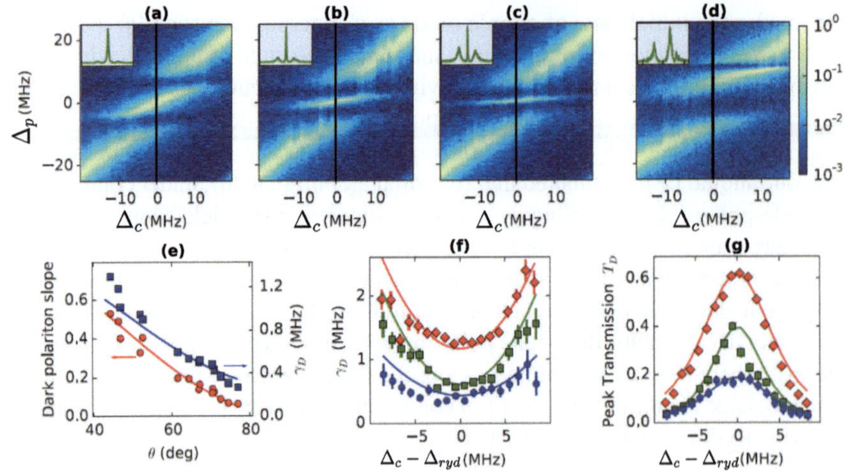

Fig. 4.8 The cavity-Rydberg polariton spectroscopy. **a–c** Transmission spectrum as a function of external control field detuning Δ_c and probe field detuning Δ_p for different polariton rotational angles θ(deg)$= 43, 62$ and 72: for different resultant transition frequencies $\Omega = 13.1(1), 6.9(1)$ and $4.9(1)$MHz; and for different atom-field couplings $G = 12.3(2), 13.0(1)$ and $14.7(1)$MHz, respectively, when $|\Delta_{ryd}| < 1$. **d** Transmission spectrum from cavity when the Rydberg blockade detuning is $\Delta_{ryd} = 9.8(4)$MHz while effective coupling and effective transitional frequency are consider $G = 16.8(3)$MHZ and $\Omega = 8.2(6)$MHz, respectively. Insect figures in **a–d** show the transmission spectrum when the intra-cavity optical detuning is set to zero ($\Delta_c = 0$) to see the behavior of EIT. Color configuration here corresponds to the strength of cavity-Rydberg transmission spectra. **e** Cavity-Rydberg polariton energy as a function of rotatory angle of dark-state polariton obtained form the polariton slope (red solid and circle curves) and damping rate γ_D, (blue solid and square curves). Here the solid curves correspond to the theoretical calculation while circle and square curves represent experimental data. **e** and **g** the influence of detuning between cavity mode and EIT resonance $\Delta_c - \Delta_{ryd}$ which cause blockade excitation in accordance to the data shown in (**a**)–(**c**). Here red, green and blue curves correspond to the results shown in (**a**), (**b**) and (**c**), respectively. Similarly, solid curves represent theoretical calculation while the other curves are for experimental results. Adopted from [69], ©(2025) with permission from American Physical Society

the dark-state polariton decay rate, in accordance to the EIT resonance, can be defined by the combination of cavity damping rate κ and decay rate of Rydberg blockaded state γ_r, as $\gamma_D \approx \kappa cos^2(\theta) + \gamma_r sin^2(\theta)$ [70]. The curve, represented with squares, in Fig. 4.8d accommodates the dependence of γ_D on the rotational angle of dark polariton along with the theoretical results obtained from the Lorentzian fit over the intra-cavity EIT resonance. It can be observe that the dark-state polariton damping rate reach to value 0.26MHz, with maximum angle, which is much larger than the line-width of the cavity, in their experiment it was about 1.8MHz, interestingly satisfying the condition of minimum quantum nonlinearity.

The polariton-polariton interactions depending upon the cavity configuration shift the intra-cavity optical mode out-off resonance with Rydberg blockade resonance, which basically causing EIT, providing a handle to detune the EIT resonance. Such intra-cavity detuning of the resonant cavity mode with resonating Rydberg-EIT provides an opportunity to enhance the lifetime of dark-state polaritons inside the cavity by decreasing their damping γ_D. This difference can be further enhance by inducing interaction between polariton excited by the multi-mode cavity field or by inducing interaction between polaritons in multiple coupled cavities driven by same external field [71–73]. However, this kind are, so-far, only possible in cavity-Rydberg system due to Rydberg blockade mechanism and in optical cavities with ground state atoms have not been studied yet [60, 74]. The damping rate of the intra-cavity polariton can be defined as a function of cavity detuning and detuning of Rydberg blockade state which causing EIT resonance, given as,

$$\gamma_D \approx \kappa cos^2(\theta) + \gamma_R sin^2(\theta) + A(\Delta_c - \Delta_{ryd}), \tag{4.24}$$

where $A = 4\Omega^2 G^2 \Gamma_{EIT}/(\Omega^2 + G^2)^3$ is the quadratic strength of the polariton damping rate define by the amplitude of the cavity-blockade detuning (Δ_{CB}) which of course will not contribute in a resonant case. Figure 4.8e illustrates, both theoretically and experimentally, such dependence of dark polariton damping as γ_D a function of cavity-blockade detuning which arises because of the off-resonant behavior of EIT with cavity mode for different rotating angles of the dark-state polariton. In order to increase dark-state polariton decay rate from cavity with condition $\gamma_R << \kappa$, it is required by the off-resonant cavity-blockade detuning to must in order of $\Delta_{CB} = \Delta_c - \Delta_{ryd} = \Omega^2/(2\Gamma_{EIT}\sqrt{D_d} cos^2(\theta))$, where $D_d = G^2/\kappa\Gamma_{EIT}$ is the polariton enhanced depth of the intra-cavity optical potential while Γ_{EIT} is the line-width of the EIT window. One can note that when the cavity-blockade detuning is tune to the frequency of the cavity mode, the energy of the dark-state of the polariton will just shift about $\Delta_{CB} cos^2(\theta) = \Omega_2/2\Gamma_{EIT}\sqrt{D_d}$ which is equal to the EIT description in free-space.

The dynamics of the dark-state polariton can be further observed and explain by measuring the cavity-Rydberg EIT bandwidth in the maximum cavity transmission spectrum, as illustrated in Fig. 4.8f. In [69], they observe that such maximum cavity-Rydberg peak transmission of the outgoing optical mode following squared Lorentzian pattern defined as,

$$T_D \propto \left[1 + \frac{(\Delta_c - \Delta_{ryd})^2}{(\Gamma'_{EIT}/2)^2}\right], \tag{4.25}$$

where, $\Gamma'_{EIT} \approx 2\Delta_{CB}$ corresponds to the peak width of EIT transmission in the case of off-resonant cavity-blockade detuning. In Fig. 4.8e, it is illustrated that the peak transmission width can be enhance with increase in rotatory angle of the cavity-Rydberg polariton. The peak width of EIT window Γ'_{EIT} also determines the polariton-polariton interaction strength which finally reveals the bandwidth for dark-state polariton in EIT. It is also responsible for the kinetic energy of dark-state polariton in a multi-mode cavity field. This peak EIT bandwidth is crucially important in implementation of quantum computation because their strength determines the interaction time in quantum logic gates. Further, they studied the sensitively of inhomogeneous broadening of the intra-cavity Rydberg atomic excitations, which arise from the Doppler effects caused by the probe field wavelength mismatch with the cavity field in JC ladder configuration (particularly, in presence cavity dielectric mirrors), to the cavity dark-state polariton.

The also discussed the occurrence of large scales electric polarization in intra-cavity optical mode after interacting with Rydberg atoms due to the excitation of dark-state polaritons. Interestingly, the decoherence of the outgoing EIT mode due to the broadening effects are independent of the blockaded state of the Rydberg atoms. These features are significantly important to coherently manipulate the optical spin waves Duan-Lukin-Cirac-Zoller type single-photons sources [44, 50]. The external control field provides this opportunity to coherently maneuver the spinning optical mode which eventually results in the suppression of polariton mode [75]. Recently, in another work, the dynamics of cavity-Rydberg polariton have been discussed by considering long-range dipole-dipole interaction regime [76]. They have found, by similarly considering two types of intra-cavity interactions including dipole-dipole interaction and interactions arising from the coherent atomic ensemble, that the intra-cavity atomic localization induced photon bunching excite a two-polariton combined state even in the presence of repulsive mechanism between different system degrees of freedom. They obtain that photon back-bunching from mismatch occurring between the quantization volume of atomic mode and intra-cavity optical mode. They also demonstrated the emergence of such bi-polariton states because of the saturability effects, or because of the coherent quantum interference, between atomic transitional pathways. The cavity-Rydberg dark-state polariton are, thus, extremely important to test and implement quantum nonlinear phenomena of modern physics.

References

1. O. Firstenberg, C.S. Adams, S. Hofferberth, Nonlinear quantum optics mediated by Rydberg interactions. J. Phys. B: At. Mol. Opt. Phys. **49**, 152003 (2016)
2. J.M. Raimond, M. Brune, S. Haroche, Manipulating quantum entanglement with atoms and photons in a cavity. Rev. Mod. Phys. **73**, 565 (2001)
3. M. Fleischhauer, A. Imamoglu, J.P. Marangos, Electromagnetically induced transparency: optics in coherent media. Rev. Mod. Phys. **77**, 633 (2005)
4. L.V. Hau, S.E. Harris, Z. Dutton, C.H. Behroozi, Light speed reduction to 17 m per second in an ultracold atomic gas. Nature **397**, 594–598 (1999)
5. K. Hammerer, A.S. Sørensen, E.S. Polzik, Quantum interface between light and atomic ensembles. Rev. Mod. Phys. **82**, 1041 (2010)
6. J. Gea-Banacloche, Y.Q. Li, S.Z. Jin, M. Xiao, Electromagnetically induced transparency in ladder-type in homogeneously broadened media: theory and experiment. Phys. Rev. A **51**, 576 (1995)
7. A.K. Mohapatra et al., A giant electro-optic effect using polarizable dark states. Nat. Phys. **4**, 890–894 (2008)
8. D. Petrosyan, J. Otterbach, M. Fleischhauer, Electromagnetically induced transparency with Rydberg atoms. Phys. Rev. Lett. **107**, 213601 (2011)
9. A.V. Gorshkov et al., Photon-photon interactions via Rydberg blockade. Phys. Rev. Lett. **107**, 133602 (2011)

10. S. Sevincli, N. Henkel, C. Ates, T. Pohl, Nonlocal nonlinear optics in cold Rydberg gases. Phys. Rev. Lett. **107**, 153001 (2011)
11. J.D. Pritchard et al., Cooperative atom-light interaction in a blockaded Rydberg ensemble. Phys. Rev. Lett. **105**, 193603 (2010)
12. D.E. Chang, V. Vuletic′, M.D. Lukin, Quantum nonlinear optics—photon by photon. Nat. Photonics **8**, 685–694 (2014)
13. T. Peyronel et al., Quantum nonlinear optics with single photons enabled by strongly interacting atoms. Nature **488**, 57–60 (2012)
14. R. Boddeda et al., Rydberg-induced optical nonlinearities from a cold atomic ensemble trapped inside a cavity. J. Phys. B: At. Mol. Opt. Phys. **49**, 084005 (2016)
15. K.A. Yasir, W.M. Liu, Controlled electromagnetically induced transparency and fano resonances in hybrid BEC-optomechanics. Sci. Rep. **6**, 22651 (2016)
16. V. Parigi et al., Observation and measurement of interaction-induced dispersive optical nonlinearities in an ensemble of cold Rydberg atoms. Phys. Rev. Lett. **109**, 233602 (2012)
17. O. Firstenberg et al., Attractive photons in a quantum nonlinear medium. Nature **502**, 71–75 (2013)
18. J. McKeever et al., State-insensitive cooling and trapping of single atoms in an optical cavity. Phys. Rev. Lett. **90**, 133602 (2003)
19. J. Stanojevic et al., Dispersive optical nonlinearities in a Rydberg electromagnetically-induced-transparency medium. Phys. Rev. A **88**, 053845 (2013)
20. M.F. Maghrebi et al., Coulomb bound states of strongly interacting photons. Phys. Rev. Lett. **115**, 123601 (2015)
21. G. Günter et al., Observing the dynamics of dipole-mediated energy transport by interaction-enhanced imaging. Science **342**, 954–956 (2013)
22. A. Grankin et al., Quantum-optical nonlinearities induced by Rydberg–Rydberg interactions: a perturbative approach. Phys. Rev. A **92**, 043841 (2015)
23. A. Imamoḡlu, H. Schmidt, G. Woods, M. Deutsch, Strongly interacting photons in a nonlinear cavity. Phys. Rev. Lett. **79**, 1467 (1997)
24. K.M. Birnbaum et al., Photon blockade in an optical cavity with one trapped atom. Nature **436**, 87–90 (2005)
25. A. Miranowicz et al., Tunable multiphonon blockade in coupled nanomechanical resonators. Phys. Rev. A **93**, 013808 (2016)
26. X.W. Xu, A.X. Chen, Y.X. Liu, Phonon blockade in a nanomechanical resonator resonantly coupled to a qubit. Phys. Rev. A **94**, 063853 (2016)
27. J.F. Huang, J.Q. Liao, C.P. Sun, Photon blockade induced by atoms with Rydberg coupling. Phys. Rev. A **87**, 023822 (2013)
28. C. Guerlin, E. Brion, T. Esslinger, K. Mølmer, Cavity quantum electrodynamics with a Rydberg-blocked atomic ensemble. Phys. Rev. A **82**, 053832 (2010)
29. S. Baur, D. Tiarks, G. Rempe, S. Dürr, Single-photon switch based on Rydberg blockade. Phys. Rev. Lett. **112**, 073901 (2014)
30. G.W. Lin et al., Strong photon blockade with intracavity electromagnetically induced transparency in a blockaded Rydberg ensemble. Phys. Rev. A **92**, 043842 (2015)
31. E. Urban et al., Observation of Rydberg blockade between two atoms. Nat. Phys. **5**, 110–114 (2009)
32. A. Gaëtan et al., Observation of collective excitation of two individual atoms in the Rydberg blockade regime. Nat. Phys. **5**, 115–118 (2009)
33. Y.O. Dudin, A. Kuzmich, Strongly interacting Rydberg excitations of a cold atomic gas. Science **336**, 887 (2012)
34. V. Vuletic, When superatoms talk photons. Nat. Phys. **2**, 801 (2006)
35. T.C. Liebisch, A. Reinhard, P.R. Berman, G. Raithel, Atom counting statistics in ensembles of interacting Rydberg atoms. Phys. Rev. Lett. **95**, 253002 (2005)
36. S. Kumar et al., Collective state synthesis in an optical cavity using Rydberg atom dipole blockade. J. Phys. B: At. Mol. Opt. Phys. **49**, 064014 (2016)

37. A. Grankin et al., Inelastic photon scattering via the intracavity Rydberg blockade. Phys. Rev. Lett. **117**, 253602 (2016)
38. J. Lim, H.G. Lee, J. Ahn, Review of cold Rydberg atoms and their applications. J. Korean Phys. Soc. **63**, 4 (2013)
39. M.D. Lukin, Colloquium: trapping and manipulating photon states in atomic ensembles. Rev. Mod. Phys. **75**, 457 (2003)
40. I. Carusotto, C. Ciuti, Quantum fluids of light. Rev. Mod. Phys. **85**, 299 (2013)
41. J.M. Zhang, W.M. Liu, D.L. Zhou, Mean-field dynamics of a Bose Josephson junction in an optical cavity. Phys. Rev. A **78**, 043618 (2008)
42. H. Deng, H. Haug, Y. Yamamoto, Exciton-polariton Bose-Einstein condensation. Rev. Mod. Phys. **82**, 1489 (2010)
43. R. Balili et al., Bose-Einstein condensation of microcavity polaritons in a trap. Science **316**, 1007–1010 (2007)
44. T. Chanelie′re et al., Storage and retrieval of single photons transmitted between remote quantum memories. Nature **438**, 833–836 (2005)
45. J. Simon et al., Interfacing collective atomic excitations and single photons. Phys. Rev. Lett. **98**, 183601 (2007)
46. A.V. Akimov et al., Generation of single optical plasmons in metallic nanowires coupled to quantum dots. Nature **450**, 402–406 (2007)
47. D. Maxwell et al., Storage and control of optical photons using Rydberg polaritons. Phys. Rev. Lett. **110**, 103001 (2013)
48. D. Tiarks et al., Single-photon transistor using a Förster resonance. Phys. Rev. Lett. **113**, 053602 (2014)
49. H. Gorniaczyk et al., Single-photon transistor mediated by interstate Rydberg interactions. Phys. Rev. Lett. **113**, 053601 (2014)
50. L.-M. Duan, M.D. Lukin, J.I. Cirac, P. Zoller, Long-distance quantum communication with atomic ensembles and linear optics. Nature **414**, 413–418 (2001)
51. J. Klaers, J. Schmitt, F. Vewinger, M. Weitz, Bose-Einstein condensation of photons in an optical microcavity. Nature **468**, 545–548 (2010)
52. D.E. Chang et al., Crystallization of strongly interacting photons in a nonlinear optical fibre. Nat. Phys. **4**, 884 (2008)
53. N. Schine et al., Synthetic Landau levels for photons. Nature **534**, 671–675 (2016)
54. G. Liu, N. Hao, S.L. Zhu, W.M. Liu, Topological superfluid transition induced by a periodically driven optical lattice. Phys. Rev. A **86**, 013639 (2012)
55. F. Sun et al., Topological quantum phase transition in synthetic non-abelian gauge potential: gauge invariance and experimental detections. Sci. Rep. **3**, 2119 (2013)
56. X.L. Zhang, L.F. Liu, W.M. Liu, Quantum anomalous hall effect and tunable topological states in 3d transition metals doped silicene. Sci. Rep. **3**, 290 (2013)
57. K. Jachymski, P. Bienias, H.P. Büchler, Three-body interaction of Rydberg slow-light polaritons. Phys. Rev. Lett. **117**, 053601 (2016)
58. M.J. Gullans et al., Effective field theory for Rydberg polaritons. Phys. Rev. Lett. **117**, 113601 (2016)
59. F. Ripka, Y.H. Chen, R. Löw, T. Pfau, Rydberg polaritons in a thermal vapor. Phys. Rev. A **93**, 053429 (2016)
60. H. Wu, J.G. Banacloche, M. Xiao, Observation of intracavity electromagnetically induced transparency and polariton resonances in a doppler-broadened medium. Phys. Rev. Lett. **100**, 173602 (2008)
61. J. M. Me′nard et al., Revealing the dark side of a bright exciton-polariton condensate. Nat. Commun. **5**, 4648 (2014)
62. G. Hernandez, J. Zhang, Y. Zhu, Vacuum Rabi splitting and intracavity dark state in a cavity-atom system. Phys. Rev. A **76**, 053814 (2007)
63. O.A. Egorov, D.V. Skryabin, A.V. Yulin, F. Lederer, Bright cavity polariton solitons. Phys. Rev. Lett. **102**, 153904 (2009)

64. M. Fleischhauer, M.D. Lukin, Dark-state polaritons in electromagnetically induced transparency. Phys. Rev. Lett. **84**, 5094 (2000)
65. M.D. Eisaman et al., Electromagnetically induced transparency with tunable single-photon pulses. Nature **438**, 837–841 (2005)
66. T. Peyronel et al., Quantum nonlinear optics with single photons enabled by strongly interacting atoms. Nature **488**, 57–60 (2012)
67. W. Chen et al., All-optical switch and transistor gated by one stored photon. Science **341**, 768–770 (2013)
68. G.W. Lin et al., Optical cavity quantum electrodynamics with dark-state polaritons. Phys. Rev. A **89**, 043815 (2014)
69. J. Ningyuan et al., Observation and characterization of cavity Rydberg polaritons. Phys. Rev. A **93**, 041802(R) (2016)
70. M.D. Lukin, M. Fleischhauer, M.O. Scully, V.L. Velichansky, Intracavity electromagnetically induced transparency. Opt. Lett. **23**(4), 295–297 (1998)
71. A.C. Ji, X.C. Xie, W.M. Liu, Quantum magnetic dynamics of polarized light in arrays of microcavities. Phys. Rev. Lett. **99**, 183602 (2007)
72. A.C. Ji, Q. Sun, X.C. Xie, W.M. Liu, Josephson effect for photons in two weakly linked microcavities. Phys. Rev. Lett. **102**, 023602 (2009)
73. L. Tan, Y.Q. Zhang, W.M. Liu, Quantum phase transitions for two coupled cavities with dipole-interaction atoms. Phys. Rev. A **84**, 063816 (2011)
74. M. Mücke et al., Electromagnetically induced transparency with single atoms in a cavity. Nature **465**, 755–758 (2010)
75. A. Amo et al., Exciton-polariton spin switches. Nat. Photonics **4**, 361–366 (2010)
76. M. Litinskaya, E. Tignone, G. Pupillo, Cavity polaritons with Rydberg blockade and long-range interactions. J. Phys. B: At. Mol. Opt. Phys. **49**, 164006 (2016)

Chapter 5
Cavity Mediated Rydberg Quantum Phase Transitions

Abstract This chapter focuses on collective many-body phenomena arising in cavity-coupled Rydberg ensembles. The superradiant phase transition of cavity-Rydberg polaritons is discussed, along with the enhancement of superradiance by strong RydbergRydberg interactions. The analysis extends to ordered phases in two-dimensional Rydberg arrays and phase transitions in cold-atom lattices. Particular emphasis is placed on the interplay of local versus nonlocal quantum dynamics, string dynamics, and emergent correlations, which highlight the rich nonequilibrium physics accessible in these systems.

5.1 Introduction

The transformation of matter into different phases—such as gas, liquid, solid, and, under extreme conditions, plasma–based on variations in temperature and pressure is a fundamental physical process known as a *phase transition*, specifically a *classical phase transition*. These transitions are typically characterized by discontinuous changes in the thermodynamic variables of a system and are described using the principles of classical thermodynamics. For instance, the transition of water from a liquid phase to a solid phase at the freezing point, or from a liquid to a gaseous phase at the boiling point, are examples of first-order classical phase transitions. These phenomena occur due to thermal fluctuations and involve latent heat and symmetry breaking associated with macroscopic order parameters.

In contrast, when temperature approaches absolute zero ($T = 0$ K), thermal fluctuations become negligible, and the transition between different phases of matter is governed by purely quantum mechanical effects. Such transitions are referred to as *quantum phase transitions* (QPTs). These are fundamentally different from their classical counterparts because they are driven by quantum fluctuations stemming from the Heisenberg uncertainty principle, rather than thermal agitation. Quantum phase transitions typically occur by tuning a non-thermal control parameter such as magnetic field, pressure, or interaction strength in a Hamiltonian system. They are characterized by a fundamental change in the ground state of the system and

K. A. Yasir and W.-M. Liu, *Rydberg Atoms in Cavity*, Springer Series in Solid-State Sciences 213, https://doi.org/10.1007/978-981-95-4571-1_5

can be classified as second-order (continuous) transitions associated with diverging correlation lengths and critical behavior [1].

QPTs are of profound importance in understanding exotic states of matter and many-body quantum phenomena. They provide insights into the emergence of quantum order, criticality, and universality classes in strongly correlated systems. In particular, they form the theoretical underpinning for understanding complex behavior in systems ranging from magnetic insulators and superconductors to quantum Hall fluids and topological materials. The study of QPTs has found a natural testbed in *quantum many-body hybrid systems*, including cold atoms in optical lattices, trapped ions, and superconducting circuits, where precise control over interactions, dimensionality, and external fields is possible.

Over the last decade, the rapid advancement in the experimental realization of ultra-cold atoms, especially in the strongly interacting regime, has significantly enhanced our ability to explore and manipulate quantum phase transitions in controllable settings. For example, the transition from a Mott-insulating phase to a superfluid phase in a Bose-Hubbard system was one of the first direct realizations of a QPT in cold atom platforms [2–8]. Other remarkable demonstrations include the realization of a Tonks-Girardeau gas [9], the BEC-BCS crossover [10], and the Kosterlitz-Thouless transition in two-dimensional systems [11]. More recently, there has been considerable interest in exploring topologically nontrivial phases, including topological insulators and superconductors [12–15], and topological superfluids arising from strong spin-orbit coupling and pairing mechanisms.

Furthermore, experiments have demonstrated the emergence of symmetry-protected topological orders and long-range entangled phases, which lie outside the traditional Landau paradigm. These efforts have brought quantum phase transitions to the forefront of contemporary condensed matter and atomic physics, spurring extensive theoretical and experimental investigations into critical phenomena, universality classes, entanglement scaling, and out-of-equilibrium quantum dynamics [16–18]. A particularly stunning and powerful platform to explore QPTs is provided by the field of *cavity quantum electrodynamics* (cavity QED), especially when integrated with ultra-cold or Rydberg atoms. Cavity QED enables the coherent manipulation of light-matter interactions at the quantum level and allows exploration of novel quantum phases of light and matter under strong coupling conditions. When two-level atoms, such as Rydberg atoms exhibiting long-lived excited states and strong dipole-dipole interactions, are placed inside high-finesse optical cavities, they collectively interact with the quantized cavity mode, giving rise to exotic many-body phenomena, including nonlocal correlations and emergent order.

One of the hallmark quantum phases in this context is the *superradiant phase*, which emerges from the collective enhancement of spontaneous emission due to the coherent coupling of atoms with the cavity field. This phenomenon is well described by the *Dicke model*, which captures the collective interaction between a large number of two-level atoms and a single bosonic mode of the electromagnetic field [19–21]. The Dicke model predicts a phase transition between a normal phase (with zero field amplitude) and a superradiant phase (with macroscopic occupation of the cavity mode), breaking the continuous $U(1)$ symmetry associated with the atomic-field coupling. The superradiant phase is interpreted as a condensation of light and matter into a new collective state known as a *polariton*, which embodies hybridized excitations of atomic and photonic degrees of freedom. This symmetry-breaking phase transition has been realized in several platforms, including cold atom condensates and circuit QED architectures [22]. Importantly, the inclusion of Rydberg atoms in such systems introduces additional tunability due to their strong van der Waals or dipole-dipole interactions. These interactions enable the simulation of long-range interacting spin models, leading to novel phases of matter that combine both collective atom-field dynamics and interatomic correlations. Consequently, cavity QED systems with Rydberg atoms provide an exciting venue to study the interplay between coherent light-matter coupling and quantum many-body interactions. They facilitate the realization of spin-boson models, lattice gauge theories, and topological quantum simulators, thus deepening our understanding of quantum criticality, non-equilibrium phase transitions, and emergent phenomena in hybrid quantum systems.

5.2 Rydberg Superradiant-Phase Transition

5.2.1 Superradiant Phase of Cavity-Rydberg Polariton

The demonstration of Rydberg atoms in one dimensional lattice [23] has contributed a lot in the study of strongly correlated many-body system, ranging from spintronic systems [24] to the melting of highly condense systems [25]. Such superradiant phase transitions in cavity-Rydberg polariton was first studied by Zhang et al. [26], in 2013, where the they found that the Rydberg-Rydberg interaction, inside an optical cavity driven by a Rabi laser, leads to three incompressible quantum phases in Rydberg solid states corresponding to the atomic filling number 0, $1/2$ and 1, which eventually destroy the coherence of intra-cavity optical mode. Most interestingly, they discover the novel state of matter in the form of superradiant-solid phase which basically is a combination of superradiant phase and crystalline solid phase caused by the broken $U(1)$ and translational symmetries of atomic structure. In such phase, the superradiant and solid characteristics of matter coexist. The superradiant-solid phase somehow appears to the analog of super-solid phase which emerges in optical lattices except that it is caused by the condensation of the polariton by nonlocal matter-wave interactions. They also demonstrated that, in cavity-Rydberg polariton, the quantum phase transition from solid, with atomic filling number 0 and 1, to superradiant phase remain second order as compare to the phase transition from solid, with filling number $1/2$, and superradiant-solid to the superradiant phase, which appears to in first order.

They consider an optical lattice, with N' sites, lying along the ultra-high finesse cavity having optical wavelength λ_0 and single-photon Rabi coupling g_0, as shown in schematic Fig. 5.1a. Each site of the lattice is coupled to a Rydberg atom which has a ground state $|g\rangle$ and a highly excited Rydberg state $|e\rangle$. First, the intra-cavity optical mode excite atoms from ground state to the intermediate state $|p\rangle$, with detuning Δ_p, and then external Rabi laser will excite atom from central state $|p\rangle$ to the Rydberg

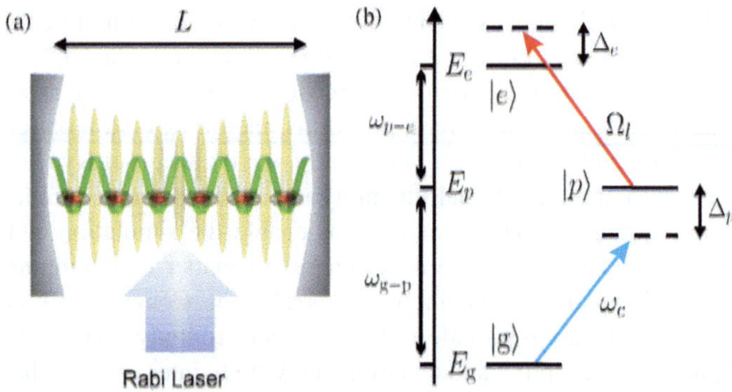

Fig. 5.1 **a** Schematic diagram for cavity-Rydberg system where the Rydberg atoms are trapped in a one dimensional optical lattice inside a high-finesse cavity of length L and driven by single mode Rabi laser ω_l. Each site of the lattice contains one Rydberg atom having ground state $|g\rangle$ and highly excited Rydberg state $|e\rangle$ defined by two-photon transition configuration. **b** The level configuration for a Rydberg atoms trapped in single site of lattice. Intra-cavity optical mode triggers the transition of a Rydberg atom from ground state $|g\rangle$ (E_g) to an intermediate state $|p\rangle$ (E_p) with detuning $\Delta_p = \omega_c - \omega_{g-p}$, where ω_{g-p} is the transition frequency from ground state to the intermediate state. The external Rabi field triggers the transition of a Rydberg atom from intermediate state $|p\rangle$ to the excited Rydberg state $|e\rangle$ (E_e) with detuning $\Delta_e = \omega_l - \omega_{p-e}$, where ω_{p-e} is that transitional frequency. Obtained from [26], ©(2025) with permission from American Physical Society

state $|e\rangle$, with detuning $\Delta_e = \omega_l - \omega_{p-e}$ defined by the frequency of the Rabi laser and the transitional frequency from state $|p\rangle$ to $|e\rangle$, as shown in the Fig. 5.1b. If we consider the detuning Δ_p very large then the intermediate state $|p\rangle$ can be ignore and the system will be transformed to the two-level Rydberg blockade and the effective Hamiltonian for each site coupled to an atom by following JCM, under rotating-wave approximation, can be written as,

$$\hat{H}_i^{eff} = -\Delta|e_i\rangle\langle e_i| + g(\hat{\psi}|e_i\rangle\langle g_i| + H.c.), \tag{5.1}$$

where $\Delta = \Delta_p + \Delta_e$ is the total detuning of the system and operator $\hat{\psi}^\dagger$ $(\hat{\psi})$ corresponds to the creation (annihilation) operator of the effective cavity field. $g = g_0^* \Omega_l / |\Delta_p|$ is the coupling of the atom to the lattice site depending on the single-induced electric field effects. By apply unitary transformation under a new rotating frame and by considering two-photon process for intra-cavity one dimensional optical lattice, one can drive the effective Hamiltonian for whole system [26],

$$\hat{H}_{eff} = \omega_c \hat{\psi}^\dagger \hat{\psi} + \sum_{i=1}^{N} \frac{\epsilon}{2}\left(b_i^\dagger b_i - a_i^\dagger a_i\right) + g \sum_{i-1}^{N} \left(b_i^\dagger a_i \hat{\psi} + H.c.\right)$$
$$+ V \sum_{\langle ij \rangle} P^{(i)} P^{(j)} - \mu N_{ex}, \tag{5.2}$$

where a_i and b_i are the bosonic field operators corresponding to the atomic ground state $|g_i\rangle$ and excited state $|e_i\rangle$ in the lattice site i obeying the single occupancy condition $b_i^\dagger b_i + a_i^\dagger a_i = 1$. The atom-field coupling can be assumed to be uniform inside cavity by simply taking the optical lattice spacing equal to the wavelength of the cavity mode. To reveal the strong blockaded polaritons, they considered scaled atom-field coupled as $g = \bar{g}/\sqrt{N}$. In order obtain strong Rydberg-Rydberg interaction between two nearby Rydberg atoms, they map the projector operator $P^{(i)} = (1 + \sigma_z^{(i)})/2$ on the effective Rydberg state with transitional amplitude $\sigma_z^{(i)} = b_i^\dagger b_i - a_i^\dagger a_i$ which defines the nearest neighbor interaction strength in order of $V = C_6/a^6$ depending upon the principal quantum number. The last term in Hamiltonian (5.2) accommodates the chemical potential as a function of total number of atomic excitation with intra-cavity field, $N_{ex} = \hat{\psi}^\dagger\hat{\psi} + \sum_{i=1}^{N} b_i^\dagger b_i$. If we consider the Rydberg-Rydberg interactions zero ($V = 0$) in above Hamiltonian, which is explainable by the Dicke mode, then the atomic ground state $|g\rangle$ is not appear to be directly coupled to the excited Rydberg state $|e\rangle$ which consequently leads to a scenario where the no-go theorem [27] is not further applicable which gives the chance of superradiant phase.

Further, if we consider the atom-field coupling zero ($g = 0$), which converts the system into a pure lattice of Rydberg atoms, then by manipulating chemical potential over the cavity field $\tilde{\mu} = \mu - \omega_c$, one can reveal the three different types of Rydberg solid phases depending upon the conditions: first, when $\tilde{\mu} < -\Delta$, the all Rydberg atoms then stay in the ground state which lead to a solid-0 phase (i.e. solid phase with filling number 0); second, when $-\Delta < \tilde{\mu} < 2V - \Delta$, then the half of the Rydberg atoms stay in ground state while the other half will be excited to the Rydberg state $|e\rangle$ which cause the occurrence of solid$-1/2$ phase; finally, when $\tilde{\mu} > 2V - \Delta$, then the solid$-1$ phase occurs because the all atoms are now in excited Rydberg state.

Such scenario, where the intra-cavity atom-field interaction and the repulsion between Rydberg atoms are competing with each other, provides an opportunity to break $U(1)$ and translational symmetries simultaneously by cooling the combine state of light and matter in the form of polariton which may result in a new quantum phase of superradiant-solid. Generally, it is difficult to achieve both $U(1)$ and translational broken at the same time but due to cavity mediated nonlocal atom-field interactions and long-range Rydberg-Rydberg interaction provide that opportunity and make possible to reveal new phase of matter. In [26], to observe superradiant-solid phase numerically, they apply the variational-ground state wave function to the system, defined as,

$$|\lambda, \theta\rangle = exp\left(\frac{\lambda\sqrt{N}\hat{\psi}^\dagger}{2}\right)\prod_i \left[cos(\frac{\theta_i}{2})b_i^\dagger + sin(\frac{\theta_i}{2})a_i^\dagger\right]|0\rangle, \tag{5.3}$$

where $|0\rangle$ corresponds to the intra-cavity initial vacuum state when all Rydberg atoms are in their ground state. The parameters λ and θ_i represent variational parameters obtained from the variational wave function and are associated with the coherent intra-cavity optical mode and atomic field, respectively, which intuitively connected to the dynamics of cavity-Rydberg polariton.

In order to obtain the atom-light polariton in the ground state, it is necessary to calculated the energy density $\mathcal{E}$ over these variational parameters by using effective system Hamiltonian,

$$\begin{aligned}\mathcal{E} &= 4\langle \lambda, \theta | \hat{H}_{eff} | \theta, \lambda \rangle \\ &= 4\Big[-\bar{g}\lambda(sin\theta_C + sin\theta_D) + Vcos\theta_C cos\theta_D - \tilde{\mu}\lambda^2 \\ &\quad -(\tilde{\mu} + \Delta - V)(cos\theta_C + cos\theta_D)\Big] \end{aligned} \tag{5.4}$$

where subscript C and D are associated with the two subsequent sub-lattices angles which basically reveals the dominated components of the solid phase with filling number $1/2$. In above Eq. (5.4), the effective coupling parameter $\bar{g}$ provides a handle to enhance the intra-cavity lattice parameter λ which can consequently lead to condensation of cavity-Rydberg polariton with a favorable staggering order defined by the parameter V. The phase diagram, obtain from the minimized energy density over the variational parameters and representing different phase of cavity-Rydberg polariton, is shown in Fig. 5.2 and the dependence of different solid phases on the variational

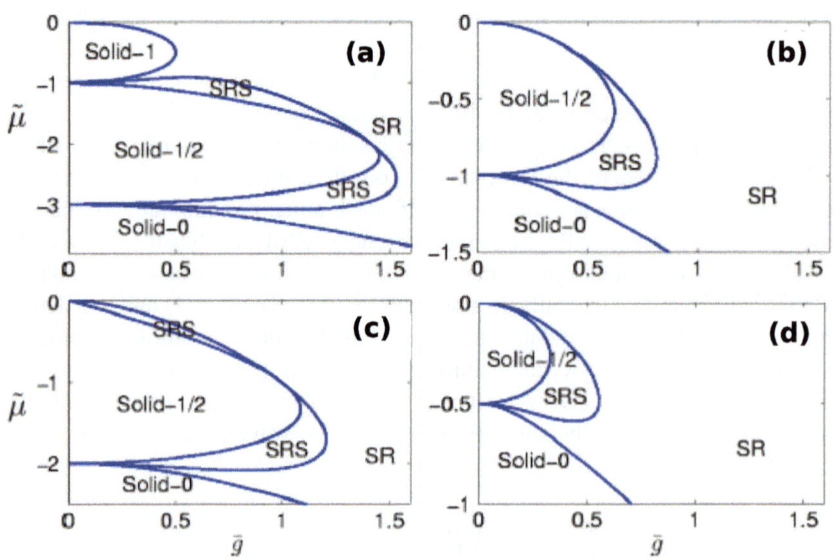

Fig. 5.2 **a–d** The cavity-Rydberg quantum phase diagram corresponding to the variational parameters calculated from Eq. (5.4) and obtained by plotting $\tilde{\mu}$ versus g for different system detuning strengths $\Delta = 3, 2, 1$ and 0.5, respectively. SR corresponds to superradiant region while SRS is for the superradiant-solid phase. All parameters used are scaled by the parameter V. Adopted from [26], ©(2025) with permission from American Physical Society

Table 5.1 The configuration of variational parameters for different phase transitions under the condition $\theta_1 \neq \theta_2$ for superradiant-solid phase. These variational values are obtained from Eq. (5.4) which is used to calculate the phase diagram shown in Fig. 5.2. Adopted from [26]

Variational values	Solid-0	SRS	SR	Solid-1/2	Solid-1
θ_C	π	θ_1	θ_{SR}	0	0
θ_D	π	θ_2	θ_{SR}	π	0
λ	0	λ_{SRS}	λ_{SR}	0	0

parameters are shown in Table 5.1, where the emerging solid phase corresponds to the Rydberg crystalline structure independent to the cavity mode excitations.

The region indicated with SR represents the superradiant phase where the condensation of the atom-light polariton occurs with the combined effects of nonzero order parameter $\langle b_i^\dagger a_i \rangle = sin\theta_{SR}/2$ and the intra-cavity optical mode parameter λ_{SR}, here subscript SR are for particular variational parameters for superradiant phase as shown in Table 5.1. On the other hand, region indicated with $Solid - 0$, $Solid - 1/2$ and $Solid - 1$ correspond to the quantum solid phases with filling number 0, 1/2 and 1, respectively. One can note the emergence of another quantum phase, when the effective atom-field coupling is approximately equal to the parameter V, of cavity-Rydberg polariton between the solid phase with filling 1/2 and superradiant phase, indicated by the SRS, which can be called as a superradiant-solid phase. In superradiant-solid phase, not only the $U(1)$ symmetry breaks but it also breaks the translational symmetry, unlike superradiant phase where the atoms are excited uniformly inside cavity, because of the atomic excitation density $\rho_{CD} = (cos\theta_{1/2})/2$ appears to be not equal in two sub-lattices as it should be in translational symmetry. Therefore, it consequently results in a combined form of crystalline phase and a superradiant phase, which is a superradiant-solid phase. Further, Fig. 5.2 also shows the effects of effective system detuning on the dynamics of quantum phase transition in cavity-Rydberg polariton and, thus, the cavity-Rydberg system not only facilitates in investigating different new quantum phases but effective system detuning also provides an opportunity to control these phase transitions.

As these results, presented in [26], are calculated under the umbrella of quantum mean-field theory, therefore, the superradiant-solid phase may not appears to be stable when the effects of nonlinear quantum fluctuation are considered. Especially, when more (quasi-particles) or less (quasi-holes) excitations of Rydberg states are consider during the solid-1/2 phase which may result in very strong quantum fluctuations from the nonlocal intra-cavity atom-field interaction and excite free quasi-particles (or holes) on the intra-cavity optical lattice. This will consequently destroy the staggering order in the solid phase which will force solid-1/2 phase to directly shift to superradiant phase and we will not get superradiant-solid phase. Therefore, it is necessary to calculate unbiased system numerically in order to obtained stable superradiant-solid phase. For this purpose, they [26] adopted the Stochastic Cluster Series Expansion (SCSE) method [26, 28] to obtain the highly accurate results of quantum phase transitions. SCSE possess a crucial importance in strongly correlated many-body system because it can effectively and efficiently simulate the dynamics of such system in presence of strong and long-range interactions. In order to perfectly distinguish these quantum phase, one most have to calculate many system observables; including average of the atomic excitation density $\rho = \langle N_{ex}\rangle/N$, structure factor (or the dynamics structure factor [29, 30]) $S(Q) = \langle|\sum_{k=1}^{N} n_k e^{iQr_k}|^2\rangle/N$ and the compressibility corresponding to the elastic factor of the matter $\kappa_T = N\beta(\langle\rho^2\rangle - \langle\rho\rangle^2)$. Particularly, structure factor occupy a key role in observing and understanding occurrence of superradiant-solid phase, where, by considering quality factor $Q = \pi$, one can realize the broken translational symmetries, as demonstrated in [26] and illustrated in Fig. 5.3 As the order parameter $\langle b_i^\dagger a_i\rangle$ and structure factor $S(Q)$ are connected to each other and cause the occurrence of sixth order terms in the free-energy expression, therefore, superradiant-solid phase emergences in first order phase transitions which is unlike to the other phase transition.

Figure 5.3 illustrates normalized structure factor $S(Q)/N$ and compressibility $\kappa_T/10$, as a function of $\tilde{\mu} - \tilde{\mu}_c$, corresponding to the trajectories where superradiant-

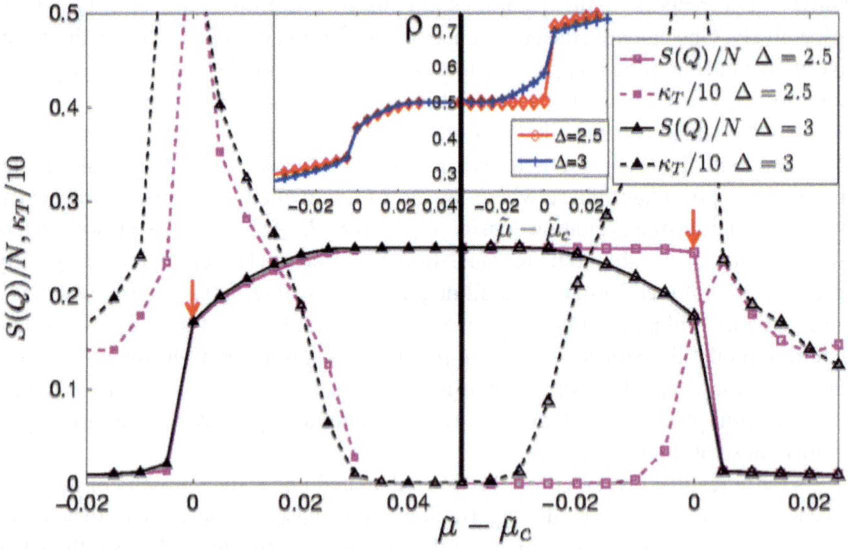

Fig. 5.3 The normalized structure factor $S(Q)/N$ (solid with square and triangle curves) and compressibility factor $\kappa_T/10$ (dashed with square and triangle curves) as a function of $\tilde{\mu} - \tilde{\mu}_c$. Here the left panel corresponds to the upper quantum phase transition point from superradiant to superradiant-solid phase while the right panel is for the lower transition point, as shown in the phase diagram Fig. 5.2. The pink square curves correspond to the effective system detuning $\Delta = 2.5$ while black triangle curves represent detuning $\Delta = 3$. The red arrows represent the upper and lower phase transition points. Here $V = 1$, $\beta = 500$ and $N = 100$. The inset graph illustrates the behavior of average atomic excitation density ρ as a function of $\tilde{\mu} - \tilde{\mu}_c$. Here the red curve corresponds to system detuning $\Delta = 2.5$ and the blue curve illustrates ρ at $\Delta = 3$. Adopted from [26], ©(2025) with permission from American Physical Society

solid phase transition occurs from superradiant phase [26], as indicated by the red arrow in the graph. To minimize the thermal effects, a normalized temperature parameter is considered $T = V/500$ for the typical intra-cavity lattice formed by the $N = 100$ atomic sites. One can note in both left and right panels, which basically indicates upper and lower superradiant phase transitions respectively, the emergence of the zero structure factor $S(Q)/N \to 0$ together with the $\kappa_T/10$, which reveals that the compressibility still holds even in the presence of translational symmetry causing the occurrence of superradiant phase. Besides, where compressibility approaches to zero $\kappa_T \to 0$ with nonzero structure factor $S(Q)/N \neq 0$ corresponding to the solid phase with half filling and, interestingly, the region where both structure factor and compressibility are nonzero corresponds to the transition of superradiant-solid phase because translational symmetries no longer holds. In that particular region, structure factor $S(Q)$ along with the average atomic excitation density ρ, as illustrated by the inset figure, shows a finite step corresponding to the phase transition while compressibility κ_T illustrates an unsaturated behavior corresponding to the critical value of $\tilde{\mu} - \tilde{\mu}_c$ which leads to the phase transition.

The effects of effective system detuning are also illustrated in Fig. 5.3, where one can note that the detuning Δ effects for the upper phase transition panel are almost negligible but for the lower transition panel both structure factor and compressibility get shrinked and decay faster with the increase in system detuning. It is because of the second-order light-matter interaction which leads to a mismatch between second-order energies of particle and hole, defined as $E_p \sim \bar{g}^2/(\Delta - 2C_6)$ and $E_h \sim \bar{g}^2/\Delta$, respectively. These quantum descriptions of different phases of cavity-Rydberg polariton have not only contributed in understanding various forms of matter but also have a lot of contributions in the field of cavity quantum electrodynamics and, recently, a lot of investigations have been performed in this regard, including the study of quantum phases in propagating light [31], the dynamics of exciton and polariton in an intra-cavity optical lattice [32] and a comparison between short and long-range many-body interactions in an optical lattice with the help of quantum phase transitions [33].

5.2.2 *Interaction-Enhanced Superradiance*

The concept of superradiance, introduced by Dicke in 1954, describes the phenomenon where an ensemble of initially uncorrelated quantum emitters spontaneously develops phase coherence and emits light collectively, leading to a radiation intensity that scales as N^2, where N is the number of emitters. This cooperative effect emerges from indistinguishability and collective coupling to a common electromagnetic mode, typically treated under the Dicke model. While originally studied in idealized atomic ensembles, superradiance has since been realized in a variety of systems including trapped ions, quantum dots, superconducting qubits, and cold atoms in optical cavities. In the traditional Dicke model and its open-system extensions, such as the dissipative Dicke model, the superradiant phase transition occurs when the collective atom-cavity coupling exceeds a critical threshold. However, these models generally assume that the emitters are either non-interacting or only weakly coupled via mediated exchange through the cavity field. The situation changes dramatically when emitters possess strong, direct interactions—as is the case in Rydberg atom arrays–raising the question: How do strong internal correlations within the atomic medium modify the threshold and nature of the superradiant phase transition? The recent work by Yiwen et al. [34] addresses this question by considering a cavity QED system composed of an array of strongly interacting Rydberg atoms coupled to a single-mode microwave cavity. The key result is that the presence of repulsive Rydberg interactions can enhance the superradiant phase, even allowing it to emerge at arbitrarily small atom-cavity coupling strengths under specific conditions. This discovery not only extends the conventional understanding of superradiance but also reveals a new regime of interaction-enabled coherent light generation in open quantum systems.

The system under investigation consists of a one-dimensional array of ultracold atoms, each individually trapped in optical tweezers or lattice sites with subwavelength spacing. The atoms are coherently driven to high-lying Rydberg states, resulting in an effective two-level system for each atom: a ground state $|g\rangle$ and an excited Rydberg state $|r\rangle$, Fig. 5.4 (upper panel). The atoms are placed inside or near a superconducting microwave cavity such that they couple collectively to a single quantized cavity mode with frequency ω_c. An external coherent drive with frequency ω_d stimulates transitions between $|g\rangle$ and $|r\rangle$, mediated via the cavity field. The geometry and driving conditions are chosen such that all atoms couple homogeneously to the cavity field with a real coupling constant g. At the same time, due to their Rydberg character, the excited states experience strong, long-range van der Waals interactions characterized by a pairwise potential $V_{jk} = C_6/|r_j - r_k|^6$. In most experimental realizations, the nearest-neighbor interaction dominates due to the rapid decay of the C_6 potential, but the full long-range character can still have nontrivial effects.

In the frame rotating with the drive frequency ω_d, the system's Hamiltonian is given by:

$$\hat{H} = -\Delta_c \hat{a}^\dagger \hat{a} - \sum_{j=1}^{N} \Delta \hat{\sigma}_j^+ \hat{\sigma}_j^- + g \sum_{j=1}^{N} \left(\hat{a} \hat{\sigma}_j^+ + \hat{a}^\dagger \hat{\sigma}_j^- \right) + \sum_{j<k} V_{jk} \hat{\sigma}_j^+ \hat{\sigma}_j^- \hat{\sigma}_k^+ \hat{\sigma}_k^- , \tag{5.5}$$

where $\hat{a}$ is the annihilation operator for the cavity photons, and $\hat{\sigma}_j^\pm$ are the raising and lowering operators for the j-th atom, acting on the Rydberg transition. The detunings are defined as $\Delta = \omega_d - \omega_0$ and $\Delta_c = \omega_d - \omega_c$ for the atoms and cavity, respectively.

The system is inherently open, with cavity photon decay occurring at rate κ, and spontaneous emission from the Rydberg states occurring at rate γ. The corresponding master equation is:

$$\frac{d\hat{\rho}}{dt} = -i[\hat{H}, \hat{\rho}] + \kappa \mathcal{D}[\hat{a}]\hat{\rho} + \gamma \sum_j \mathcal{D}[\hat{\sigma}_j^-]\hat{\rho}. \tag{5.6}$$

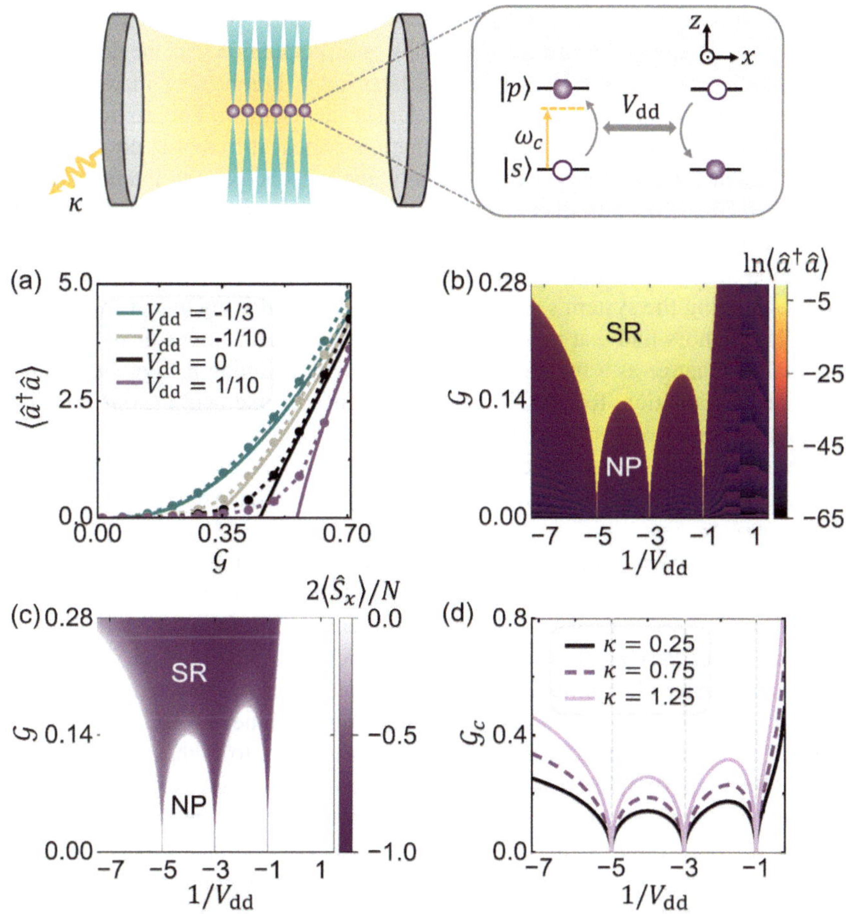

Fig. 5.4 The upper panel illustrates the system schematic setup. While lower panel discusses interaction-enhanced superradiance and the role of Rydberg interactions. **a** Steady-state cavity photon number $\langle \hat{a}^\dagger \hat{a} \rangle$ versus atom-cavity coupling strength g, shown for various values of Rydberg interaction strength V_0. With increasing V_0, the critical coupling g_c for superradiance decreases and eventually vanishes at certain discrete interaction values, indicating that strong interatomic interactions can induce superradiance even in the absence of external driving. **b** Phase diagram showing the dependence of the critical coupling g_c on V_0. The nonmonotonic behavior and zeros of g_c identify specific interaction strengths where interaction-induced superradiance emerges. **c** Susceptibility $\chi = \partial \langle \hat{a}^\dagger \hat{a} \rangle / \partial g$ or rescaled photon number near the critical points, demonstrating diverging response and mean-field-like criticality associated with the interaction-induced phase transitions. **d** Time evolution of the cavity photon number $\langle \hat{a}^\dagger(t) \hat{a}(t) \rangle$, showing the dynamical onset of the superradiant phase. For strong enough interactions (large V_0), the cavity field builds up over time even when the coupling g is weak, confirming the spontaneous emergence of macroscopic coherence. Overall, the panels collectively demonstrate that many-body interactions in a driven-dissipative Rydberg system can drastically alter the superradiant phase boundary and dynamically stabilize coherent light emission. Adopted from [34], ©(2025) with permission from American Physical Society

To make analytical and numerical progress, the authors consider the mean-field limit where all atoms are treated identically, and fluctuations are neglected. This approximation is justified for sufficiently large atom numbers or in cases where fluctuations are suppressed by dissipation. Under the mean-field approximation, the expectation values of relevant operators obey nonlinear differential equations that can be solved for the steady-state. Crucially, the presence of the interaction term $\sum_{j<k} V_{jk}\hat{\sigma}_j^+\hat{\sigma}_j^-\hat{\sigma}_k^+\hat{\sigma}_k^-$ introduces a nonlinear shift in the energy of the Rydberg state that depends on the population of neighboring atoms. This acts effectively as a population-dependent detuning, modifying the resonance condition for each atom and thus altering the system's susceptibility to coherent driving and collective coupling. The authors find that the steady-state cavity photon number $\langle\hat{a}^\dagger\hat{a}\rangle$ exhibits a non-analytic change as a function of the atom-cavity coupling g, signaling a dissipative phase transition. Remarkably, in the presence of interactions, this transition occurs at progressively smaller values of g, until at certain discrete values of the interaction strength V_0, the critical coupling g_c vanishes. At these points, the system spontaneously organizes into a superradiant steady state even in the limit of vanishing cavity coupling.

In Fig. 5.4(lower panel), the authors present the core numerical and analytical results that establish the phenomenon of interaction-enhanced superradiance. The subfigures Fig. 5.4a–d, each highlighting a different aspect of the superradiant phase transition under varying Rydberg interaction strength. The Fig. 5.4a displays the steady-state cavity photon number $\langle\hat{a}^\dagger\hat{a}\rangle$ as a function of the single-atom coupling strength g, for several fixed values of the interaction parameter V_0. Each curve represents a different V_0, increasing from lowest at the top to highest at the bottom (or vice versa depending on labeling). In the non-interacting case ($V_0 = 0$), there is a clear threshold g_c above which the photon number jumps from nearly zero to a macroscopic value, signifying the onset of the conventional dissipative Dicke superradiant phase. As V_0 increases, the threshold g_c decreases systematically, and at certain critical interaction strengths V_0^*, the photon number becomes non-zero even as $g \to 0$. This behavior directly demonstrates that sufficiently strong Rydberg interactions can induce a superradiant state without requiring any appreciable atom–cavity coupling. The Fig. 5.4b maps the phase boundary by plotting the critical coupling g_c as a function of the Rydberg interaction strength V_0. The curve declines toward zero at discrete points V_0^*, indicating where the critical coupling vanishes. Between those points, g_c remains finite, defining a normal-to-superradiant boundary. The non-monotonic structure–namely, the presence of multiple zeros–reflects the quantized nature of collective atomic states entering into degeneracy at specific interaction strengths, as described by emergent quantum Rabi models in the supplemental theory. The diverging susceptibility near these degeneracies underlies the dramatic enhancement of superradiance.

Figure 5.4c typically shows either a rescaled plot indicating universal scaling of photon number near the critical points or (in some versions) the diverging susceptibility $\chi = \partial\langle\hat{a}^\dagger\hat{a}\rangle/\partial g$ evaluated near g_c, as a function of V_0. This subfigure emphasizes that near the discrete critical interaction values, even infinitesimal driving or coupling produces a large photon response, consistent with a mean-field-type non-equilibrium phase transition with diverging slope and susceptibility. Such scaling data support the interpretation that the system undergoes a second-order dissipative critical transition at each V_0^*. The Fig. 5.4d illustrates the time-dependent dynamics of the cavity photon number $\langle \hat{a}^\dagger(t)\hat{a}(t) \rangle$ for different values of the Rydberg interaction strength V_0 at a fixed atom-cavity coupling g. The results show that in the absence of interactions ($V_0 = 0$), the cavity photon number remains negligible over time, indicating no buildup of coherent radiation. However, as the interaction strength increases, especially near the discrete critical values of V_0 identified in the phase diagram Fig. 5.4b, the cavity field exhibits a rapid increase in photon number, converging to a steady superradiant value even when g is relatively small. This behavior confirms that the Rydberg-mediated interactions can dynamically drive the system into a superradiant phase, without the need for strong external coupling. The emergence of a macroscopic photon population from an initially vacuum-like state highlights the critical role of long-range dipolar interactions in the nonequilibrium stabilization of light-matter coherence.

This interaction-enhanced superradiance can be understood as a cooperative effect where long-range repulsion between Rydberg excitations leads to an effective stabilization of the excited state population against decay. As the excitation of one atom suppresses the excitation of nearby atoms (the so-called Rydberg blockade), the system avoids saturation and maintains a favorable population inversion, allowing for the buildup of macroscopic coherence. Moreover, the cavity field acts as a mediator of long-range coherence, amplifying the effect of the atomic interactions. The combination of these effects results in a hybrid light-matter steady state where the atoms are phase-coherent and the cavity contains a macroscopic photon population. Such a state is inherently nonequilibrium, stabilized by a balance between coherent driving, dissipation, and interactions. The discovery of interaction-enhanced superradiance enriches the landscape of nonequilibrium quantum phase transitions. It demonstrates that the interplay between coherent many-body interactions and dissipation can lead to entirely new forms of quantum order, inaccessible in equilibrium settings. This phenomenon shares similarities with driven-dissipative phase transitions seen in polariton condensates, spin-boson lattices, and open quantum Ising models, and adds a new layer of tunability via Rydberg physics. From an experimental perspective, the proposed system is highly feasible using current technology. The techniques to trap and address Rydberg atom arrays with single-site resolution have been developed in pioneering works such as that by Bernien et al. [35], and coupling such arrays to superconducting microwave cavities is a subject of growing interest. These platforms could allow direct observation of interaction-tuned superradiant transitions, as well as exploration of related phenomena such as bistability, quantum criticality, and time crystals in open Rydberg systems. Finally, the theoretical framework developed in this study paves the way for generalizations to higher-dimensional arrays, disorder-induced localization effects, and fully quantum treatments beyond mean-field. Understanding how quantum fluctuations and entanglement manifest in such driven, interacting systems remains a major challenge and opportunity in modern quantum optics.

5.3 Two-Dimensional Rydberg Array

The experimental system consists of a two-dimensional square array of ultracold Rydberg atoms trapped in optical tweezers and coupled to a high-finesse optical cavity, as illustrated in Fig. 5.5a. Each tweezer site contains exactly one atom that can be excited from the ground state $|g\rangle$ to a Rydberg state $|e\rangle$ through a two-photon transition via an intermediate state $|m\rangle$. The key feature of this setup is the coupling between the $|m\rangle \leftrightarrow |g\rangle$ transition and a single mode of the optical cavity, described by the photon creation operator $\psi^\dagger$ and annihilation operator ψ. The Rydberg atoms interact through strong dipole-dipole interactions with strength V, which decays as $1/r^6$ with distance r, making nearest-neighbor interactions dominant in the square lattice configuration.

The cavity is carefully tuned such that its resonance frequency ω_c is close to but detuned from the atomic transition frequency ϵ between $|m\rangle$ and $|g\rangle$, with the detuning given by $\Delta = \omega_c - \epsilon$. The system is driven by an external laser field that facilitates the two-photon transition, and the effective atom-photon coupling strength is denoted by g. This coupling strength g depends on both the cavity parameters and the Rabi frequency of the laser driving the $|e\rangle \leftrightarrow |m\rangle$ transition. The combination of strong, long-range dipole-dipole interactions and cavity-mediated interactions creates a rich platform for studying quantum phase transitions in a hybrid light-matter system.

The section offers a comprehensive account of the hybrid many-body system studied by An et al. [36], involving a two-dimensional lattice of Rydberg atoms embedded

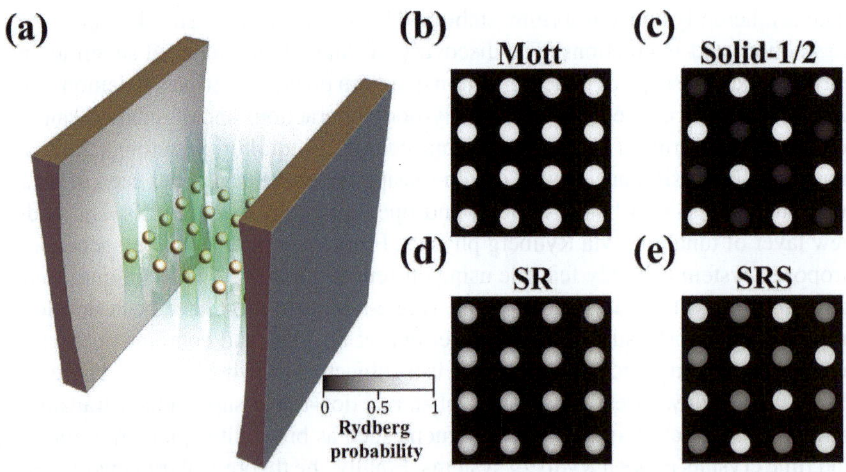

Fig. 5.5 **a** Schematic of the experimental setup showing a 2D Rydberg atom array coupled to an optical cavity. **b–e** Configurations of Rydberg excitations (red circles) in different quantum phases: **b** Mott-1 phase with all atoms excited, **c** solid-1/2 phase with checkerboard pattern, **d** superradiant phase with uniform excitation and photon coherence, and **(e)** superradiant solid phase combining checkerboard order with photon coherence. Adopted from [36], ©(2025) with permission from American Physical Society

in an optical cavity. We discuss the physical model, central Hamiltonians, and the multi-order quantum phase transitions revealed through variational theory and large-scale quantum Monte Carlo (QMC) simulations. The presentation is enriched with mathematics and referenced figures for clarity. The study of quantum phase transitions in strongly correlated systems coupled to optical cavities represents a frontier in quantum many-body physics, combining concepts from condensed matter physics and quantum optics. These hybrid light-matter systems offer unique opportunities to explore novel quantum phenomena that emerge from the interplay between long-range photon-mediated interactions and direct atomic correlations. The particular system of Rydberg atoms in optical cavities has attracted significant attention due to the exceptional tunability of Rydberg-Rydberg interactions and the strong coupling achievable in high-finesse cavities. This combination enables the exploration of quantum phases with no direct analogues in conventional condensed matter systems, where the breaking of multiple symmetries can lead to exotic states of matter.

The fundamental model under investigation combines two key quantum optical systems into a single hybrid platform: a two-dimensional array of Rydberg atoms coupled to a high-finesse optical cavity. Figure 5.5a provides a schematic representation of this system, showing several essential components working in concert. The system consists of ultracold atoms arranged in a perfect square lattice configuration, with each atom trapped by individual optical tweezers at typical spacings of 5–10 μm. These atoms can exist in either the ground state $|g\rangle$ or a highly excited Rydberg state $|e\rangle$, with principal quantum numbers typically in the range of 50–100, which gives them extraordinary properties including strong, long-range interactions.

The optical cavity forms the second crucial component of the system, supporting a single photon mode with frequency ω_c that couples to the atomic transition between an intermediate state $|m\rangle$ and the ground state $|g\rangle$. This coupling creates an effective two-level system described by the raising operator $b_i = |e\rangle_i \langle g|_i$ and lowering operator $b_i = |g\rangle_i \langle e|_i$ at each lattice site i. The cavity photons mediate long-range interactions between the atoms while the direct dipole-dipole interactions between Rydberg atoms provide strong, tunable nearest-neighbor couplings. The interplay between these two interaction mechanisms—one mediated by photons and one direct between atoms - gives rise to the rich phase diagram observed in the system.

The remaining panels of Fig. 5.5 illustrate the characteristic spatial configurations of the different quantum phases. Figure 5.5b shows the Mott-1 phase where all atoms are uniformly excited to the Rydberg state, creating a fully occupied lattice. In contrast, Fig. 5.5c depicts the solid-1/2 phase with its distinctive checkerboard pattern of Rydberg excitations, where atoms in one sublattice are excited while their nearest neighbors remain in the ground state. Figure 5.5d presents the superradiant phase, characterized by a uniform density of Rydberg excitations across all sites combined with a macroscopic photon population in the cavity. Most remarkably, Fig. 5.5e shows the superradiant solid phase that combines features of both the solid and superradiant phases—maintaining the checkerboard pattern of excitations while also supporting a coherent photon field. This dual order parameter makes the superradiant solid phase particularly interesting as it simultaneously breaks both translational symmetry (through the spatial pattern of excitations) and U(1) gauge symmetry (through the photon coherence).

The key theoretical framework for understanding these systems combines elements of the Dicke model, describing collective light-matter coupling, with extended Hubbard-type interactions characteristic of Rydberg systems. Previous work in one-dimensional configurations revealed the existence of a fragile superradiant solid phase, but its experimental observation remained challenging due to strong quantum fluctuations in low dimensions. The extension to two-dimensional systems, as presented in this work, provides crucial advantages: enhanced stability of ordered phases against fluctuations, richer symmetry breaking patterns, and closer connection to potential experimental realizations using state-of-the-art tweezer arrays. The dimensional crossover from 1D to 2D represents not just a quantitative change in system parameters but enables qualitatively new phenomena to emerge from the increased connectivity and modified density of states.

The system is governed by the Hamiltonian:

$$H = \frac{g}{\sqrt{N}} \sum_{i=1}^{N} (b^{\dagger} i\psi + \text{H.c.}) + V \sum \langle i, j \rangle n_i n_j - \Delta \sum_{i=1}^{N} n_i - \mu N_i, \tag{5.7}$$

where $b^{\dagger}i$ and bi are the creation and annihilation operators for a Rydberg excitation at site i, treated as hard-core boson operators due to the constraint that each tweezer site can host at most one Rydberg excitation. The operator $n_i = b^{\dagger}ibi$ represents the Rydberg density at site i, and $\langle i, j \rangle$ indicates summation over nearest-neighbor pairs. The parameter μ is the chemical potential that controls the total number of excitations $N_i = \psi^{\dagger}\psi + \sum_i n_i$, which is a conserved quantity in the system.

The Hamiltonian exhibits U(1) symmetry, corresponding to the transformation $\psi^{\dagger} \to \psi^{\dagger} e^{i\theta}$ and $b_i^{\dagger} \to b_i^{\dagger} e^{-i\theta}$. This symmetry can be spontaneously broken when the system enters a superradiant phase. The interplay between the cavity-mediated interaction (first term) and the direct dipole-dipole interaction between Rydberg atoms (second term) leads to a competition that drives the system through various quantum phases. The third term represents the detuning energy, and the fourth term with chemical potential μ ensures conservation of the total number of excitations. To analyze the ground state properties of the system, a variational approach is employed using the ansatz wavefunction:

$$|\lambda, \theta\rangle = e^{\lambda\sqrt{N}\psi^{\dagger}/2} \prod_i \left[\cos\left(\frac{\theta_i}{2}\right) b^{\dagger} i + \sin\left(\frac{\theta i}{2}\right) \right] |0\rangle, \tag{5.8}$$

where λ characterizes the photon coherence and θ_i represents the angle parameterizing the Rydberg excitation at site i. For systems exhibiting Néel order, the parameters θ_i alternate between two values θ_C and θ_D on the two sublattices of the square lattice, breaking translational symmetry. The energy per site is calculated from this ansatz and minimized with respect to the variational parameters λ, θ_C, and θ_D to determine the ground state properties and phase boundaries.

The quantum Monte Carlo simulations provide crucial insights beyond the variational predictions, offering a comprehensive understanding of the nature of phase transitions in this hybrid light-matter system. As evident in Fig. 5.6, the detailed evolution of order parameters reveals distinct signatures characterizing each quantum phase and the transitions between them. In the solid-1/2 phase (observed at small coupling strength g), the structure factor S(Q)/N approaches its theoretical maximum value of 0.25, corresponding to perfect antiferromagnetic Néel order in the equivalent spin language. This maximal value reflects the perfect alternation of Rydberg excitations between the two sublattices of the square array. Simultaneously, the strictly zero compressibility κ confirms the incompressible nature of this phase, where the number of Rydberg excitations is quantized and robust against small variations of the chemical potential μ.

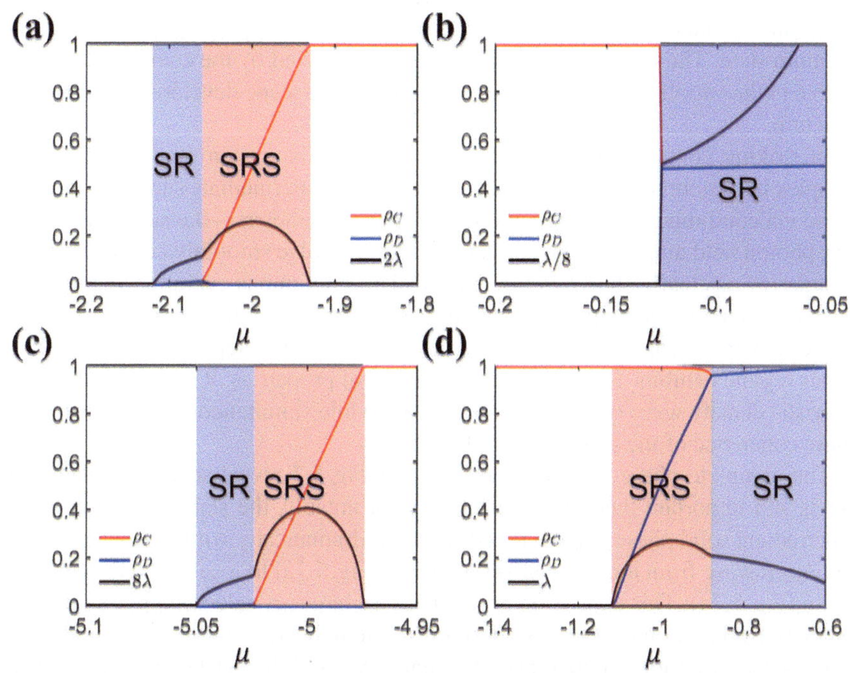

Fig. 5.6 Variational results for order parameters as functions of chemical potential μ at fixed coupling $g = 0.5$. **a, b** For $\Delta = 2$, showing first-order transitions between phases. **c, d** For $\Delta = 5$, showing continuous transitions through the SRS phase. Top panels show Rydberg densities ρ_C, ρ_D on two sublattices, bottom panels show photon order parameter λ. Colored regions indicate superradiant (blue) and SRS (red) phases. Obtained from [36], ©(2025) with permission from American Physical Society

The transition from the solid-1/2 phase to the superradiant solid phase manifests through two simultaneous and interconnected effects that are clearly visible in Fig. 5.6. First, the structure factor decreases from its maximal value of 0.25 due to the introduction of quantum fluctuations mediated by the cavity photons. Second, the compressibility becomes finite, indicating the system's newly acquired ability to support density fluctuations. This dual evolution signals the emergence of a compressible quantum phase that nevertheless maintains broken translational symmetry. The photon order parameter λ, shown in the lower panels of Fig. 5.6, simultaneously becomes non-zero, confirming the breaking of U(1) symmetry and the establishment of phase coherence in the photon field.

The detailed behavior of these phase transitions shows remarkable sensitivity to the detuning parameter Δ, as demonstrated by comparing Fig. 5.6a, b ($\Delta = 2$) with Fig. 5.6c, d ($\Delta = 5$). For smaller detuning ($\Delta = 2$), the system exhibits a direct first-order transition between the superradiant and solid-1/2 phases, characterized by abrupt, discontinuous jumps in both the sublattice densities ρ_C, ρ_D and the photon order parameter λ. These discontinuities reflect the system's inability to support a stable intermediate phase at small Δ values, forcing an abrupt reconfiguration of the quantum state. The first-order nature is further confirmed by the coexistence region visible in the variational results, where the energy landscape develops a double-well structure.

In striking contrast, at larger detuning ($\Delta = 5$), the system undergoes a more complex but smoother evolution, passing through an intermediate superradiant solid phase via continuous phase transitions. All order parameters—the sublattice densities, photon field amplitude, and structure factor—evolve smoothly across the phase boundaries, as clearly shown in Fig. 5.6c, d. This continuous evolution suggests the existence of a critical endpoint in the phase diagram where the first-order transition line terminates, giving way to second-order transitions. The superradiant solid phase in this regime exhibits a remarkable coexistence of properties: it maintains the characteristic density wave order of the solid phase while simultaneously supporting the phase coherence of the superradiant phase.

The complete phase diagrams presented in Fig. 5.7 reveal several fundamental trends and important theoretical insights. Most notably, the stability region of the superradiant solid phase (shown in red) expands dramatically with increasing detuning Δ, growing from a narrow sliver at $\Delta = 2$ (Fig. 5.7a) to a substantial domain at $\Delta = 6$ (Fig. 5.7d). This expansion occurs because larger Δ values suppress quantum fluctuations that would otherwise destabilize the delicate balance between density wave order and photon coherence. The enhanced stability at large Δ has crucial experimental implications, as it makes the observation of this exotic phase more feasible in realistic systems with finite dissipation and imperfections.

The phase boundaries obtained from quantum Monte Carlo simulations (shown as dots in Fig. 5.7) systematically deviate from the variational predictions (solid lines), particularly near the multicritical points where three phases meet. These deviations are most pronounced in the vicinity of the superradiant solid phase and reflect the importance of quantum fluctuations that are properly captured by the numerical simulations but missed by the mean-field variational approach. The quantum Monte Carlo

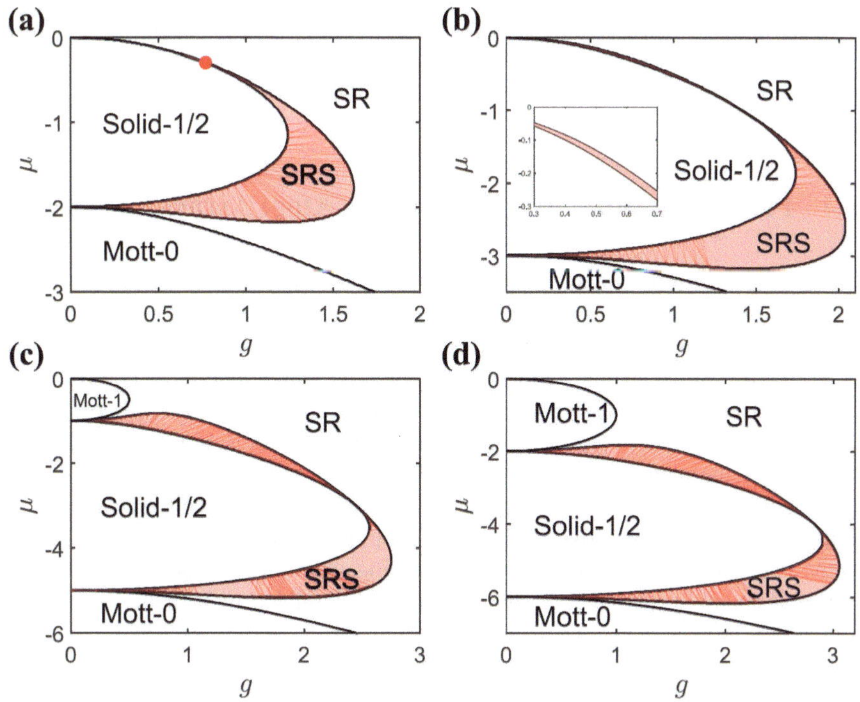

Fig. 5.7 Quantum phase diagrams in the μ-g plane for different detunings: **a** $\Delta = 2$, **b** $\Delta = 3$, **c** $\Delta = 5$, and **d** $\Delta = 6$. Red regions indicate SRS phases, with solid lines showing variational boundaries and dots marking QMC results. Insets show zoomed views of multicritical regions. Note the expansion of SRS phase with increasing Δ and emergence of particle-hole symmetry at large Δ. Adopted from [36], ©(2025) with permission from American Physical Society

results reveal that quantum fluctuations tend to suppress the extent of ordered phases, particularly in parameter regimes where competing orders are nearly degenerate in energy.

Figure 5.7 also reveals the emergence of an interesting particle-hole symmetry at large detuning (Δ = 6 in Fig. 5.7d), where the phase diagram becomes approximately symmetric about μ = - (Δ - 4V)/2. This symmetry reflects an underlying duality between excitations (particles) and vacancies (holes) in the Rydberg excitation pattern, which becomes exact in the limit of large Δ. The symmetry manifests most clearly in the symmetric arrangement of the Mott-1 and Mott-0 phases, as well as in the mirror-like quality of the phase boundaries between superradiant and solid phases.

The multicritical points visible in Fig. 5.7, particularly the triple point where superradiant, superradiant solid, and solid-1/2 phases meet, represent particularly interesting regimes for studying quantum critical phenomena. Near these points, the system's behavior becomes highly sensitive to small variations in parameters, and novel scaling behavior may emerge. The quantum Monte Carlo results suggest that the nature of these multicritical points may differ significantly from the mean-field predictions, potentially hosting unconventional critical exponents or even new universality classes specific to this light-matter hybrid system.

5.4 Phase Transition in Cold Rydberg Atom Arrays

The hybrid quantum system, in the study done by Bacciconi et al. [37], consists of a one-dimensional array of cold neutral atoms excited to Rydberg states and embedded within a high-finesse optical cavity. The atoms are trapped in optical tweezers, forming a spatially ordered configuration with single-atom control. Each atom behaves as a two-level system, with ground state $|g\rangle$ and a Rydberg-excited state $|r\rangle$, separated by a laser detuning Δ and coupled with Rabi frequency Ω. The Rydberg excitation is achieved via a two-photon transition involving an off-resonant intermediate state, and one of these photons is provided by the quantized cavity field. This gives rise to an effective atom-cavity coupling strength $g = g_0\Omega/\Delta$, where g_0 is the bare coupling rate, and Δ is the detuning from the intermediate level. This coherent coupling facilitates the emergence of hybridized light-matter excitations and allows tunable access to regimes where nonlocal cavity-mediated interactions compete with the local van der Waals interactions.

As shown in Fig. 5.8a, the atoms are driven coherently by a laser field, while also interacting via a shared cavity mode with collective light-matter coupling strength g. The cavity field mediates long-range interactions between the atoms, allowing the study of nonlocal dynamics in contrast to purely local van der Waals interactions present in free-space Rydberg ensembles. The cavity is transversely pumped and its transmission is monitored for quantum observables (Fig. 5.8d). This architecture enables a direct probe of the phase structure and excitation dynamics through the detection of transmitted light, which is sensitive to the atomic population and correlations via dispersive shifts and emission processes. The hybrid light-matter interaction forms an effective Tavis-Cummings-Ising model in one dimension, where spin-spin coupling from Rydberg blockade competes with photon-mediated all-to-all coupling.

The Hamiltonian of the system, under rotating-wave approximation, is given by:

$$\hat{H} = \sum_{i=1}^{N}\left[\frac{\Omega}{2}\hat{\sigma}_x^{(i)} - \Delta\hat{n}_i\right] + \sum_{i<j} V_{ij}\hat{n}_i\hat{n}_j + g(\hat{a}^\dagger + \hat{a})\sum_i \hat{\sigma}_x^{(i)} + \delta_c\hat{a}^\dagger\hat{a} \tag{5.9}$$

where $\hat{n}_i = |r\rangle_i \langle r|$ is the Rydberg state projector, V_{ij} is the van der Waals interaction ($\sim C_6/r_{ij}^6$), and δ_c is the cavity detuning. The coupling to the cavity introduces all-to-all interactions, making the model nontrivial and beyond mean-field solvable. The competition between the local blockade and the global photon exchange leads to a variety of quantum phases, including symmetry-breaking transitions, long-range order, and correlated many-body states. Due to the symmetry of the system, there

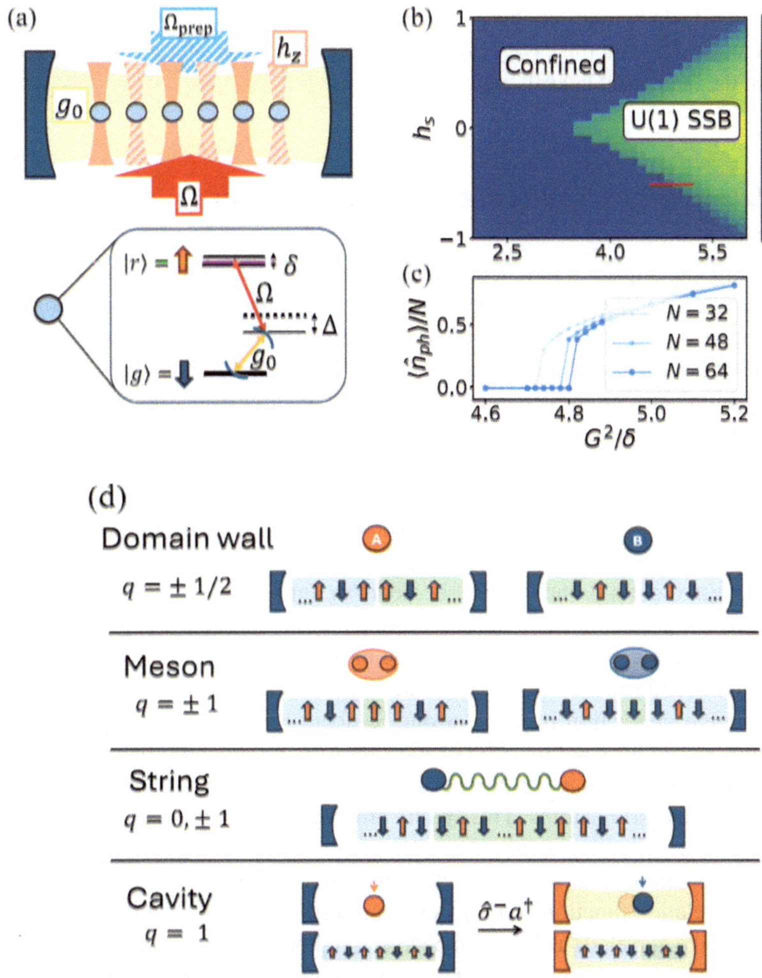

Fig. 5.8 **a** Schematic of atoms trapped in optical tweezers inside an optical cavity. **b** Atomic level structure showing Rydberg excitation with laser detuning Δ and coupling Ω, and cavity coupling g. **c** Spatial layout of atomic array with spacing a. **d** Detection setup monitoring cavity output field. Obtained from [37], ©(2025) with permission from American Physical Society

exists a global U(1) conserved charge defined by the total number of excitations in the atomic ensemble and the cavity photon number, leading to phase structures that differ from the standard Dicke-Ising class.

5.4.1 Local Versus Nonlocal Quantum Dynamics

Understanding the impact of interaction range on the collective behavior of Rydberg atom arrays is crucial for engineering and controlling many-body quantum phases. In this context, the system's dynamics are governed either by local, short-range van der Waals (vdW) interactions or by global, cavity-mediated nonlocal couplings. The competition between these two regimes leads to qualitatively different quantum dynamics and equilibrium phases. This section discusses these differences, illustrated in Fig. 5.9, which combines equilibrium observables from the quantum Monte Carlo simulations of the system. In the absence of cavity coupling, the atoms interact solely through the nearest-neighbor van der Waals interaction, which imposes a Rydberg blockade preventing two nearby atoms from being simultaneously excited. This results in spatial orderings such as checkerboard or antiferromagnetic-like patterns, depending on the geometry of the lattice and the interaction radius.

In the local regime, where the cavity coupling strength is negligible ($g \approx 0$), the dynamics and equilibrium states are dominated by the vdW interaction $V_{ij} \sim C_6/r_{ij}^6$. This short-range interaction imposes strong Rydberg blockade effects, preventing simultaneous excitations within a blockade radius. Consequently, the atoms tend to organize in a checkerboard or Néel-ordered pattern characterized by alternating excitation of neighboring atoms. This spatial ordering is evident in Fig. 5.9a, where the spin structure factor $S(q)$ exhibits a sharp peak at $q = \pi$, signaling the presence of antiferromagnetic-like correlations with a two-site periodicity. The corresponding connected correlation function $G_c(i, j)$ shown in Fig. 5.9e displays alternating positive and negative correlations between nearest neighbors, with correlations rapidly decaying beyond a few lattice sites. These signatures indicate the formation of a short-range ordered phase without long-range coherence. Such ordering is consistent with theoretical expectations from Ising-type models in one dimension, where symmetry breaking is restricted and correlations decay exponentially unless assisted by long-range interactions.

Introducing a finite cavity coupling g changes the interaction landscape dramatically. The cavity field mediates a collective, all-to-all coupling among atoms through the exchange of virtual photons, effectively acting as a quantum bus that delocalizes excitations. This nonlocal interaction competes with the local vdW forces and gives rise to emergent collective phases with more complex ordering. Figure 5.9b–d demonstrate how increasing the cavity coupling shifts and broadens the peak of $S(q)$, moving away from the simple $q = \pi$ ordering and indicating the onset of frustrated or incommensurate phases. Multiple peaks or broadened features appear, corresponding to density-wave orders with longer periodicities, such as $\mathbb{Z}_3$ or $\mathbb{Z}_4$ crystalline

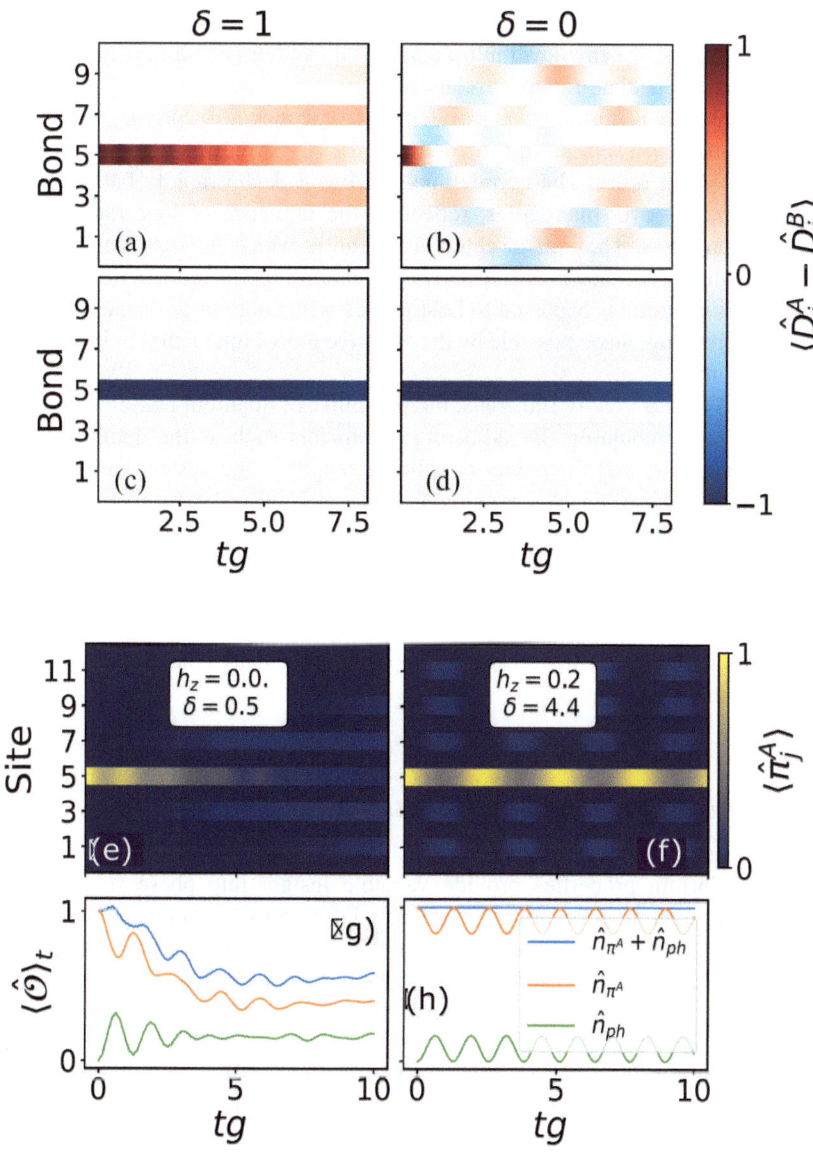

Fig. 5.9 Equilibrium properties of Rydberg atom arrays with and without cavity-mediated interactions. **a–d** Spin structure factor $S(q)$ for various detunings δ and cavity couplings, highlighting momentum-space ordering. **e–h** Connected density-density correlation functions $G_c(i, j)$ revealing spatial ordering and correlations. These results are obtained using quantum Monte Carlo simulations. Adopted from [37], ©(2025) with permission from American Physical Society

patterns. These features reflect the frustration introduced by cavity-induced long-range interactions, which allow multiple nonlocal excitation pathways and stabilize phases with intermediate-range correlations.

This transition is further corroborated by the connected correlation maps in Fig. 5.9f–h, which reveal more extended and oscillatory correlations indicative of modulated spatial order. The correlations no longer decay rapidly but show pronounced patterns over many sites, reflecting the influence of long-range cavity-mediated couplings. The system enters a regime in which the classical picture of nearest-neighbor blockade fails, and a richer quantum superposition of ordered states emerges. This regime is expected to host phases with continuous or quasi-periodic symmetry breaking, made possible by the effective global interaction induced by the cavity photons.

Overall, Fig. 5.9 vividly illustrates the tunability of quantum phases in Rydberg arrays via cavity coupling. By adjusting parameters such as the detuning δ, the Rabi frequency Ω, and the cavity coupling strength g, the system can be driven through a rich phase diagram ranging from local Néel order to exotic, nonlocal density waves and possibly glassy or frustrated phases. This highlights the power of hybrid quantum systems to simulate complex many-body phenomena that are difficult to realize in purely local models. The interplay between the coherent photon field and the atomic array introduces a platform for exploring phase transitions and non-equilibrium dynamics in the presence of long-range entanglement and hybridized quasiparticles.

5.4.2 *String Dynamics and Quantum Correlations*

While equilibrium properties provide valuable insight into phase structure, the dynamics of quantum correlations reveal how information and entanglement propagate through the system. The authors employ time-dependent variational principle (TDVP) simulations to study string excitation dynamics in two paradigmatic regimes: the short-range interacting Ising model and the cavity-coupled transverse-field Ising (TC-Ising) model. The results are presented in Fig. 5.10. The TDVP method enables tracking the evolution of domain wall excitations and correlation spreading beyond the capabilities of exact diagonalization, allowing simulation of longer time scales and larger system sizes.

Figure 5.10a and b show the time evolution of the domain wall density, a measure of the number of boundaries between spin-up and spin-down domains, which effectively tracks the spread of excitations. In the local Ising model (Fig. 5.10a), domain walls propagate ballistically, displaying a clear linear light-cone structure indicative of finite-speed quasiparticle propagation under short-range interactions. In contrast, the TC-Ising model with cavity-mediated long-range couplings (Fig. 5.10b) exhibits slower and more spatially delocalized dynamics. The cavity-induced all-to-all interaction modifies the excitation spreading by effectively correlating distant sites, which

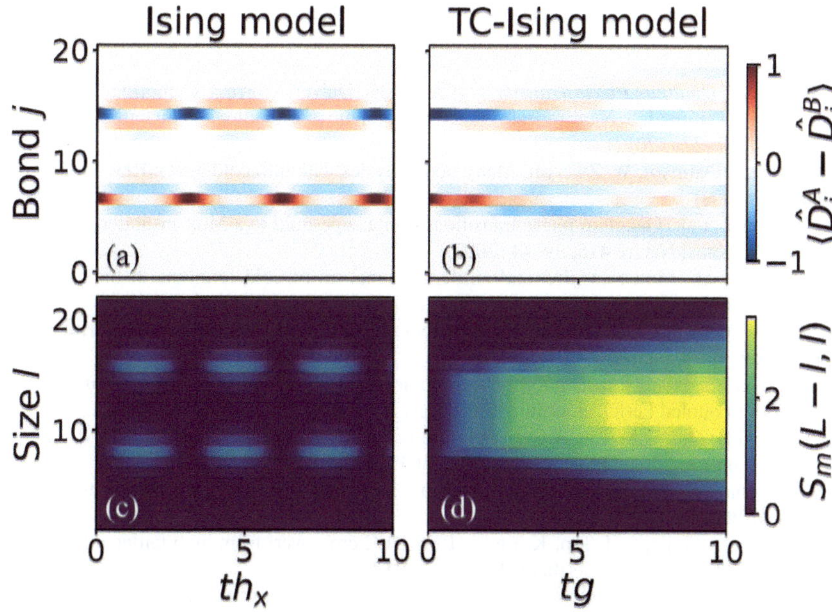

Fig. 5.10 String dynamics comparison between the Ising and TC-Ising models using TDVP simulations. **a, c**: Domain wall density and bipartite mutual information in the local Ising model with transverse field $h_x = 0.1$. **b, d**: Corresponding results in the TC-Ising model with cavity coupling $g = 0.1$ and detuning $\delta = 0$. Both models use $V = 1$, longitudinal field $h_z = 0.2$, and system size $N = 23$. These results highlight differences in the propagation of excitations and entanglement due to nonlocal cavity interactions. Adopted from [37], ©(2025) with permission from American Physical Society

results in suppression and broadening of domain wall density peaks compared to the local case. The delocalization indicates the formation of nontrivial dressed excitations, or meson-polaritons, that hybridize matter and photonic components.

The Fig. 5.10c and d depict the bipartite mutual information $S_m(L-l, l)$, quantifying total correlations between two halves of the system separated at site l. In the local Ising model (Fig. 5.10c), the mutual information grows in a nearly linear front consistent with the light-cone spreading of entanglement typical of local Hamiltonians. However, in the TC-Ising case (Fig. 5.10d), the growth is more complex and exhibits non-monotonic features with long-range correlations persisting across the system, demonstrating that the cavity mediates long-distance entanglement and modifies the causal structure of correlation propagation. These dynamical signatures reveal that even moderate cavity couplings ($g = 0.1$) dramatically alter quantum information flow and entanglement generation. The TC-Ising model's nonlocal character induces qualitatively different quantum dynamics, which could be exploited to engineer novel quantum phases and optimize quantum information protocols in cavity QED platforms.

References

1. S. Sachdev, *Quantum Phase Transitions* (Cambridge University Press, Cambridge, 1999)
2. I. Bloch, J. Dalibard, S. Nascimbene, Quantum simulations with ultracold quantum gases. Nat. Phys. **8**, 267–276 (2012)
3. I. Bloch, J. Dalibard, W. Zwerger, Many-body physics with ultracold gases. Rev. Mod. Phys. **80**, 885 (2008)
4. M. Greiner et al., Quantum phase transition from a superfluid to a Mott insulator in a gas of ultracold atoms. Nature **415**, 39–44 (2002)
5. H.F. Lin et al., Mott transition and antiferromagnetism of cold fermions in the decorated honeycomb lattice. Phys. Rev. A **90**, 053627 (2014)
6. Z.W. Xie, W.M. Liu, Superfluid-Mott-insulator transition of dipolar bosons in an optical lattice. Phys. Rev. A **70**, 045602 (2004)
7. X. Luo et al., Fidelity susceptibility and topological phase transition of a two-dimensional spin-orbit-coupled Fermi superfluid. Phys. Rev. A **89**, 043612 (2014)
8. Q. Sun, G.B. Zhu, W.M. Liu, A.C. Ji, Spin-orbit coupling effects on the superfluidity of a Fermi gas in an optical lattice. Phys. Rev. A **88**, 063637 (2013)
9. T. Kinoshita, T. Wenger, D.S. Weiss, Observation of a one-dimensional Tonks-Girardeau gas. Science **305**, 1125–1128 (2004)
10. Q.J. Chen, J. Stajic, S.N. Tan, K. Levin, BCS-BEC crossover: from high temperature superconductors to ultracold superfluids. Phys. Rep. **412**, 1 (2005)
11. Z. Hadzibabic et al., Berezinskii-Kosterlitz-Thouless crossover in a trapped atomic gas. Nature **441**, 1118–1121 (2006)
12. G. Liu, N. Hao, S.L. Zhu, W.M. Liu, Topological superfluid transition induced by a periodically driven optical lattice. Phys. Rev. A **86**, 013639 (2012)
13. G. Liu, N. Hao, S.L. Zhu, W.M. Liu, Topological superfluid transition induced by a periodically driven optical lattice. Phys. Rev. A **86**, 013639 (2012)
14. F. Sun, X.L. Yu, J. Ye, H. Fan, W.M. Liu, Topological quantum phase transition in synthetic non-abelian gauge potential: gauge invariance and experimental detections. Sci. Rep. **3**, 2119 (2013)
15. D. Zhao et al., Topological defects and inhomogeneous spin patterns induced by the quadratic Zeeman effect in spin-1 Bose-Einstein condensates. Phys. Rev. A **91**, 013619 (2015)
16. A. Bao et al., Quantum magnetic phase transition in square-octagon lattice. Sci. Rep. **4**, 6918 (2014)
17. X.L. Zhang, L.F. Liu, W.M. Liu, Quantum anomalous hall effect and tunable topological states in $3d$ transition metals doped silicene. Sci. Rep. **3**, 2908 (2013)
18. H.F. Lin, H.D. Liu, H.S. Tao, W.M. Liu, Phase transitions of the ionic Hubbard model on the honeycomb lattice. Sci. Rep. **5**, 9810 (2015)
19. R.H. Dicke, Coherence in spontaneous radiation processes. Phys. Rev. **93**, 99 (1954)
20. K. Hepp, On the superradiant phase transition for molecules in a quantized radiation field: the Dicke maser model. Ann. Phys. **76**, 360 (1973)
21. K. Baumann, C. Guerlin, F. Brennecke, T. Esslinger, Dicke quantum phase transition with a superfluid gas in an optical cavity. Nature **464**, 1301–1306 (2010)
22. P.R. Eastham, P.B. Littlewood, Bose condensation of cavity polaritons beyond the linear regime: the thermal equilibrium of a model microcavity. Phys. Rev. B **64**, 235101 (2001)
23. S.E. Anderson, K.C. Younge, G. Raithel, Trapping Rydberg atoms in an optical lattice. Phys. Rev. Lett. **107**, 263001 (2011)
24. I. Lesanovsky, Many-body spin interactions and the ground state of a dense Rydberg lattice gas. Phys. Rev. Lett. **106**, 025301 (2011)
25. H. Weimer, H. P. B*ü*chler, Two-stage melting in systems of strongly interacting Rydberg atoms. Phys. Rev. Lett. **105**, 230403 (2010)
26. X.F. Zhang et al., Rydberg polaritons in a cavity: a superradiant solid. Phys. Rev. Lett. **110**, 090402 (2013)

27. K. Rzażewski, K. Wódkiewicz, W. Żakowicz, Phase transitions, two-level atoms, and the A^2 term. Phys. Rev. Lett. **35**, 432 (1975)
28. A.W. Sandvik, Stochastic series expansion method with operator-loop update. Phys. Rev. B **59**, R14157(R) (1999)
29. K.A. Yasir, L. Zhuang, W.M. Liu, Spin-orbit-coupling-induced backaction cooling in cavity optomechanics with a Bose-Einstein condensate. Phys. Rev. A **95**, 013810 (2017)
30. K.A. Yasir, L. Zhuang, W.M. Liu, npj Quantum Inf. **8**, 109 (2022)
31. A.D. Greentree, C. Tahan, J.H. Cole, L.C.L. Hollenberg, Quantum phase transitions of light. Nat. Phys. **2**, 856–861 (2006)
32. H. Zoubi, H. Ritsch, Chapter 3—excitons and cavity polaritons for optical lattice ultracold atoms. Adv. Atom Mol. Opt. Phys. **62**, 171–229 (2013)
33. R. Landig et al., Quantum phases from competing short- and long-range interactions in an optical lattice. Nature **532**, 476–479 (2016)
34. Y. Han, H. Li, W. Yi, Interaction-enhanced superradiance of a Rydberg-atom array. Phys. Rev. Lett. **133**, 243401 (2024)
35. H. Bernien et al., Probing many-body dynamics on a 51-atom quantum simulator. Nature **551**, 579–584 (2017)
36. G.-Q. An, Y.-H. Zhou, T. Wang, X.-F. Zhang, Quantum phase transition of a two-dimensional Rydberg atom array in an optical cavity. Phys. Rev. B **106**, 134506 (2022)
37. Z. Bacciconi, H.B. Xavier, M. Marinelli, D.S. Bhakuni, M. Dalmonte, Local versus nonlocal dynamics in cavity-coupled Rydberg atom arrays. Phys. Rev. Lett. **134**, 213604 (2025)

Chapter 6
Cavity-Rydberg: Stability to Mechanics

Abstract This chapter investigates the stability and dynamical response of cavity-Rydberg systems under optomechanical coupling. Topics include optical multistability, coherent and incoherent dynamics, and the emergence of bistability in hybrid Rydbergoptomechanical systems. Mechanisms of cavity-Rydberg hybrid transduction, such as conversion between millimeter-wave and optical signals, are developed in detail. The interplay between strong atomic nonlinearities and mechanical degrees of freedom reveals novel avenues for hybrid quantum devices that integrate atomic and mechanical platforms.

6.1 Introduction

The stability of mechanical objects, which are performing harmonic oscillations or damped harmonic oscillations with respect to other surrounding degrees of freedom, has been crucially important in realistically studying and implementing many phenomena of physics. Stability deals with the particular state of a mechanical object corresponding to the variables connected to the object like position, speed, quantity, and momentum, which have a predictable value with time or space, or with any relative parameter. Inversely, there can be unstable states of the same mechanical object where the mechanical variables are highly random and unpredictable, which makes the behavior of the mechanical system chaotic. These both stable as well as unstable states of mechanical systems are determined by the system configuration and related parameters. In classical mechanics, stability criteria are often derived from potential energy landscapes and perturbative analysis, where small deviations from equilibrium states determine whether the system returns to its original configuration or diverges from it.

In quantum physics, not only stable states but also unstable states possess great importance because some of the quantum mechanical phenomena are the signature of such chaotic states, like dynamical localization where it is necessary for the system to be in a chaotic domain [1]. Such localization is a quantum interference effect

K. A. Yasir and W.-M. Liu, *Rydberg Atoms in Cavity*, Springer Series in Solid-State Sciences 213, https://doi.org/10.1007/978-981-95-4571-1_6

that arises in classically chaotic systems, and it has deep implications for wavefunction dynamics and transport properties. Especially, in highly correlated and coupled many-body systems, like cavity optomechanics, where the mechanical effects are produced by manipulating incident light in associated degrees of freedom, these chaos-induced quantum phenomena are essential and have been thoroughly investigated [2, 3]. Cavity optomechanics provides a powerful platform to study these effects by introducing nonlinear couplings between light and mechanical motion, leading to bistability, chaos, and even quantum phase transitions.

In such highly correlated systems, the mechanical objects, including intra-cavity optical modes in the form of photons, may have two (bistable) or more than two stable states (multistable) depending upon their coupling to other associated degrees of freedom. This multistability often arises from feedback between radiation pressure and cavity detuning, or from coupling with additional nonlinear elements such as atomic ensembles or superconducting qubits. The presence of multistable configurations gives rise to rich dynamical behaviors such as hysteresis, sudden switching, and noise-activated transitions. These nonlinear dynamics can be exploited for sensing, memory, or switching applications in quantum technologies.

For the cavity-Rydberg system, these stable and unstable states are even more important due to the potential applications of Rydberg atoms in various fields of quantum physics. Rydberg atoms, due to their large principal quantum number, possess enormous polarizability and long-range dipole–dipole interactions, making them highly responsive to cavity fields and external perturbations. Their integration into cavity QED platforms introduces additional interaction channels and enhances the nonlinearity of the system. The strong coupling between Rydberg atomic ensembles and cavity photons may lead to exotic dynamical regimes such as superradiance, photon blockade, and quantum phase transitions.

In this chapter, we will discuss such multiple stable and unstable states for intra-cavity Rydberg atoms due to their strong interaction with the cavity field. We will examine the underlying Hamiltonian dynamics, explore semiclassical and quantum treatments of stability, and map out the phase space trajectories under varying physical parameters. Further, we will review the recent developments in cavity optomechanics with Rydberg atoms and discuss their dynamics in strong interaction and coupling regimes. Special emphasis will be placed on how tunable interactions, collective excitations, and photon-mediated nonlinearities give rise to novel regimes of multistability and chaos, thereby enriching the control landscape for quantum simulation and quantum information processing.

6.2 Optical Multistability

Optical multistability of the cavity photons is usually referred to two or more states of the intra-cavity photon number for the single value of external field intensity. These multiple stable states are generated because of the coupling of intra-cavity optical mode with other degrees of freedom, such as cold or ultra-cold atomic gases. These multistable states can be computed by deriving the steady-state solutions for the intra-cavity photon number against the input field intensity. In the case of cavity-Rydberg system, these stable and chaotic states emerge because of the coupling of different Rydberg atoms in different states with the cavity field which trigger coherent and incoherent quantum interfaces which lead towards the predictable and unpredictable state of the cavity field. As discussed previously, Rydberg atoms are very important to analyze quantum optical nonlinearities due to their strong interaction, therefore, Rydberg atoms give us the similar advantage to demonstrate and understand such bistable dynamics of optical degree of freedom quantum mechanically. The intra-cavity optical mode bistability induced by the coupled atomic degree of freedom was first predicted by Gibbs et al. [4] in 1976, where they studied nonlinear optical dynamics by observing gain and large hysteresis in the outgoing optical mode from a Fabry-Pérot interferometer coupled to the Na vapor. After this, the optical bistable behavior has been extensively studied in different configuration by manipulating atom-field interaction and possesses great interest for the design and implementation of quantum optical filter, memory elements and many other quantum devices. However, for the cavity-Rydberg system, the optical bistability was first demonstrated by Lange et al. [5] in 1988, where they studied the optical bistability for a very low photon number coupled to a small amount of Rydberg atoms inside a superconducting microwave cavity. They observed the effects of quantum fluctuations on the bistable states of cavity photons. They experimentally illustrated the occurrence of bistable state of cavity field because of the background quantum fluctuation arising from the thermal blackbody radiations, by considering Rydberg atoms with a large principal quantum number and high quality factor $Q = 5 \times 10^6$ cavity. They also predicted that one can further enhance these background blackbody effects on the optical nonlinear dynamics by decreasing the temperature of the microwave cavity.

The multistable state of intra-cavity optical field coupled to the N Rydberg atoms passing through the cavity was demonstrated by Casagrande et al. [6] in 1993, where they injected a beam of Rydberg atoms with a limited number of atoms per pulse to a cavity driven by a single mode optical field. The advantage of the cavity-Rydberg system lye behind the long-lifetime of spontaneous emission of radiation, which consequently lead towards the broadening of intra-cavity atomic linewidth for the optical field providing chance to get strong atom-field interaction. If a two-level Rydberg atoms having excited state enters in an optical cavity, one will get an amplification of photonic signal, or get a maser effects, depending upon the resonant condition, as discussed in detail previously. However, if the two-level atom enters in a cavity with initially having ground state then the intra-cavity atom-field interaction brings no amplification and the interacting photon with atom will be absorbed. In [6], authors mainly focus on the optical bistability in absorption domain, however, they briefly discussed the emergence of optical bistability in amplification regime too, as we will discuss in this book.

To solve such cavity-Rydberg system and obtain bistable state of optical field, one can follow semiclassical theory by adopting Maxwell-Bloch approach. As the atomic broadening of transition levels is unexplainable with the help of atomic damping term, therefore, one must consider the motion of atoms while they pass through the cavity. By considering the excited atomic beam entering into a cavity along $\hat{z}$-axis, transverse to the incident optical field, and treating intra-cavity optical mode operator as a classical field variable α, the scaled dynamical Maxwell-Bloch equation for the cavity-Rydberg system, under slowly varying envelope condition, (for details see [6]) can be written as,

$$\frac{d}{dt}\alpha(t) = -k[(1 - i\Delta')\alpha(t) - \alpha_{in}] + \frac{gN}{\tau}\int_0^{\tau} dz' r^-(z', t), \tag{6.1}$$

$$\left(\frac{\partial}{\partial t} + \frac{\partial}{\partial z'}\right) r^-(z', t) = 2g\alpha(t) r_3(z', t) - i\Delta r^-(z', t), \tag{6.2}$$

$$\left(\frac{\partial}{\partial t} + \frac{\partial}{\partial z'}\right) r_3(z', t) = -g[\alpha(t) r^+(z', t) + \alpha^*(t) r^-(z', t)], \tag{6.3}$$

along with the differential equation of complex conjugates in Eqs. 6.1 and 6.2. The first equation corresponds to the intra-cavity optical mode while the other two accommodate the dynamics of atomic degree of freedom. The parameter $k = \omega_c/2Q$ is the cavity line-width defined by the intra-cavity optical field frequency ω_c and the cavity quality factor Q. The total intra-cavity photon number can be define by the classical variable for the intra-cavity field α as, $|\alpha|^2 = n$ and the parameter α_{in} is the external single mode field that drive the cavity. $\Delta = \omega_a - \omega_i n$ is the atomic transitional detuning define by the atomic frequency ω_a and the transitional frequency ω_{in} while $\Delta = \omega_c - \omega_{in}$ is the cavity field detuning. The $r^{\pm}$ are the atomic position dependent polarization variable while r_3 determines the one-half atomic population inversion. The scaled position variable for atomic mode is $z' = z/v_0$, v_0 atomic group velocity, and term $\tau = L/v_0$ is the time of flight for atoms inside cavity defines by the cavity length L.

The position and time of flight can be chosen depending upon the condition $0 \le z' \le \tau$ and the steady-state response of the system can be obtained by substituted time derivative equal to zero, as,

$$0 = -k[(1 - i\Delta')\alpha - \alpha_{in}] + \frac{gN}{\tau}\int_0^{\tau} dz' r^-(z'), \tag{6.4}$$

$$\frac{\partial}{\partial z'} r^-(z') = 2g\alpha(t) r_3(z') - i\Delta r^-(z'), \tag{6.5}$$

$$\frac{\partial}{\partial z'} r_3(z') = -g[\alpha r^+(z') + \alpha^* r^-(z')], \tag{6.6}$$

with the complex conjugates corresponding to Eqs. 6.4 and 6.5. By using the initial conditions $r^{\pm}(0)0$ and $r_3(0) = \pm 1/2$, where $\pm$ signs correspond to the absorptive and amplification configuration, respectively, of atomic mode, the solution of position dependent atomic degree of freedom will read as,

$$r_{\uparrow,\downarrow}^{-} = \pm\frac{g\alpha}{\Omega}\left[sin(\Omega z') - i\frac{\Delta}{\Omega}(1 - cos(\Omega z'))\right], \tag{6.7}$$

$$r_{3,\uparrow,\downarrow} = \pm\frac{1}{2} \mp \frac{4g^2|\alpha|^2}{\Omega^2}sin^2(\frac{\Omega z'}{2}) = \pm\frac{1}{2} \mp P_{\uparrow,\downarrow}(z'), \tag{6.8}$$

where $\uparrow$, $\downarrow$ notations corresponds to the absorption and amplification, respectively, while $\Omega = \sqrt{4g^2|\alpha|^2 + \Delta^2}$ is the effective Rabi frequency for the atomic mode passing through cavity. The parameters $P_\uparrow$ and $P_\downarrow$ represent the probability amplitudes for finding atoms in ground state and excited state, respectively.

By using atomic mode solution in Eq. (6.7) in the steady-state equation of intra-cavity optical mode (i.e. Eq. (6.4)) and by considering scaled (dimensionless) values of intra-cavity field variable $x = 2g\alpha\tau$ and input field intensity variable $y = 2g\alpha_{in}\tau$, the resultant value of input field amplitude, as a function of intra-cavity steady-state field amplitude, can be written as,

$$\begin{aligned} y = \ & x\left[1 \mp 4C\frac{1 - cos\sqrt{|x|^2 + (\Delta\tau)^2}}{|x|^2 + (\Delta\tau)^2}\right. \\ & \left. + i\left(\Delta' \pm 4C\Delta\tau\frac{1 - sin\sqrt{C^2(|x|^2 + (\Delta\tau)^2)}}{|x|^2 + (\Delta\tau)^2}\right)\right], \end{aligned} \tag{6.9}$$

where $C = Ng^2\tau/4k$ is the cooperation parameter defining the effective coupling of atomic mode with intra-cavity optical field. Further, if we consider the resonant case, means that the intra-cavity optical mode oscillates with the same frequency of atomic transitional frequency, then the steady-state solution for external driving field as a function of cavity mode will yield in [6],

$$y = x \mp C\frac{1 - cosx}{x}, \tag{6.10}$$

where the $+$ sign corresponds to the absorption setup where atoms are initially in ground state while the $-$ sign deals with amplification (stimulated emission) which yields when atoms are initially in excited state.

The stable and unstable dynamics of the intra-cavity optical degree of freedom can be illustrated in absorptive configuration by simply plotting $y = x + 4C(1 - cosx)/x$, as shown in Fig. 6.1, where the emergence multiple stable and unstable optical states can be seen. Here one should note that the dynamics of intra-cavity optical mode can be characterized depending upon the strength of external driving field. If the amplitude of input field is $y << 1$ then the cavity mode amplitude will reduced do $x \sim y/(1 + 2C)$ and if the input field amplitude is greater than 1 or C ($y >> 1, y >> C$) then the amplitude of input field and cavity mode will almost be equal $x \approx y$. Therefore, it is necessary to take strong input field in order to observe and discuss multistable dynamics of cavity field, as can be seen in Fig. 6.1. Further, the occurrence of multiple state or unstable state also depends on the cavity-Rydberg cooperativity parameter which is defined by multiple fundamental

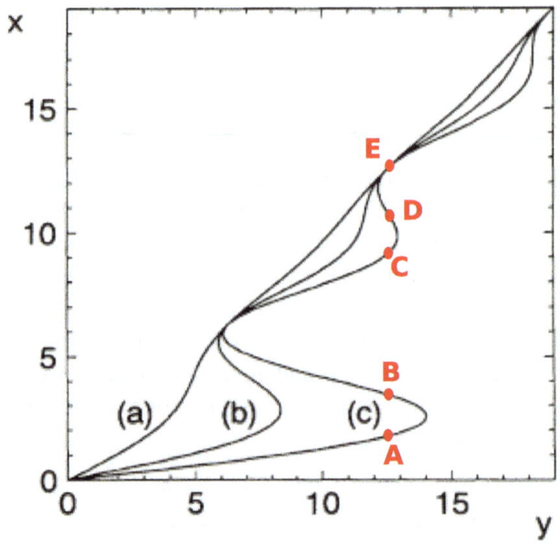

Fig. 6.1 The absorptive stable and unstable behavior of externally driven cavity mode in strong coupling the Rydberg atoms. Curves (*a*), (*b*) and (*c*) illustrate the amplitude of intra-cavity field x as a function of external driving field y when the values of cooperativity parameter are $C = 0.5$, 2 and 4, respectively. The red points (A, B and C) correspond to the stable and unstable states of the cavity field. Adopted from [6], ©(2025) with permission from American Physical Society

parameters, such as, number of atoms, intra-cavity flight time, cavity line-width and most importantly atom-field coupling. In [6], the threshold for cooperation parameter is defined by $C >> 0.85$ in order to obtain bistable or multiple stable state of cavity field. In Fig. 6.1, one can note that when cooperativity parameter is less than the threshold value (i.e. $C << 0.85$), the intra-cavity field amplitude (or intra-cavity photon number which may assumed as a cavity-Rydberg polariton state in quantum picture) always possess one value corresponding to the input field amplitude and the cavity mode is appeared to be stable for all the powers of external driving field which may referred to the monostability.

However, when we increase the value of cooperation parameter beyond the threshold (i.e. $C >> 0.85$), there will be multiple state and unstable state of cavity field corresponding to the input field, as illustrated by the curve (*b*) and (*c*) in Fig. 6.1, where the cooperation parameter is shift to $C = 2$ and $C = 4$, respectively. In curve (*b*), one can not the appearance of three solution for cavity field at a single amplitude of external input field which is called as a bistable behavior of cavity field where the cavity mode have two stable state and one unstable state, depending upon the positive and negative slop of the curve. On the other hand, curve (*c*), with $C = 4$, illustrates the occurrence of five solution of cavity field with input field which yields in tristability of the intra-cavity photons with two unstable states. The stability and unstability depend on the positive and negative slop of the curves, respectively, as (for example) illustrated by the red points in curve (*c*). The points A, C and E corresponds to the stable points, because of positive slop, where the cavity mode is stable while points B and D are the saddle-points, because of negative slop, where the cavity mode is unstable and possess chaotic behavior. Thus, one can demonstrate multistability (or unstability) in absorptive configuration of strongly driven cavity-Rydberg system by taking care of atom-field coupling and cavity characteristics which basically generate Rabi cycles yielding in multistability.

Similarly, stability analysis in amplification setup, where the Rydberg atoms are initially in excited state which leads to the stimulated emission when they interact with cavity field, can be performed by following equation $y = x - 4C(1 - cosx)/x$. It can also be characterize over the input field amplitude as [6]; if $y << 1$, then $x \approx y/(1 - 2C)$ and if $y >> 1$, then $x \approx y$. It reveals that, like absorptive setup, the external input field amplitude must be greater than one in order to observe multiple stable state of cavity field. Figure 6.2 illustrates the behavior of cavity field as a function of external input field where monostable (a), bistable (b) and tristable (c) behavior can be observe with cooperativity parameter $C = 0.5, 5$ and 12, respectively. Similar to the absorptive multistability, one can note that the cavity mode have only one stable when $C = 0.5$ and have two and three stable states for $C = 5$ and 12, respectively. Here one must note that the stability and unstability are describe by the negative and positive slop, respectively, which is unlike (inverse) to the absorptive configuration, as (for example) indicated by the red points in curve (b), where points A and C correspond to the stable points while B corresponds the saddle-points, where

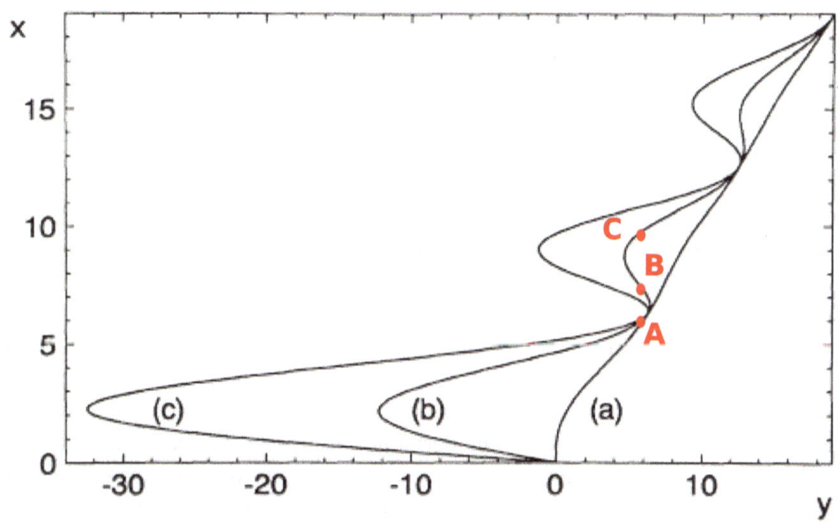

Fig. 6.2 Stability analysis of the cavity-Rydberg system in amplification regime. Here curves (a), (b) and (c) demonstrate cavity field amplitude x as a function of input field amplitude y for different values of cooperation parameter, $C = 0.5, 5$ and 12, respectively. The red dots (A, B and C) indicate the stable and unstable states of cavity mode. Adopted from [6], ©(2025) with permission from American Physical Society

cavity mode is chaotic. Further, they [6] also discuss the effects of multistability on the cavity-Rydberg maser both in absorptive and amplification setup and demonstrated the effects of atomic velocity distribution, both thermal and Gaussian, and broadening of cavity spectrum on the multistable dynamics of cavity mode.

Such nonlinear dynamics of cavity-Rydberg system stood in the foundations of many stunning quantum phenomena, including quantum nonlinear optics, optical switching and quantum gates, which crucially important to realistically implement modern quantum computation and communication. Further, cavity-Rydberg optical stability analysis was discussed in [7], where the authors not only discuss the occurrence of multiple stable and bistable states, by following same formalism as in [6], but also demonstrated the temporal instability and self-pulsing in the absorptive domain. They discussed that the self-pulsing occurs when the intra-cavity atomic mode is correlated to the multiple state states of the cavity field, especially when the atom-field coupling is larger than the atomic transitional frequency. Further such nonlinear dynamics of cavity-Rydberg system have also been discuss in [8], where authors predicted the intra-cavity optical bistability for the photon number of strongly driven cavity field as a function of atomic mode density distribution and recently in [9], where authors demonstrated multiple optical switching in cavity-Rydberg system by following same machanism as in [6].

6.3 Cavity-Optomechanics with Rydberg Atoms

The field of optomechanics deals with the manipulation of mechanical effects of light, in the form of radiation pressure, to couple electromagnetic field with the motional component of physical objects [10, 11]. The effects of radiation pressure was first observe by Kepler in the beginning 17th century [12]. He discovered the shift in the dust-trail of a passing by comet due to the radiation pressure of the Sun light. After this, the first experimental illustration of radiation pressure force of electromagnetic fields, as predicted by Maxwell, was performed by utilizing light-mill configuration [13]. Further, Einstein revealed the fluctuating and frictional nature of radiation pressure [14], which basically force them to mimic as a movable optical mirror, and this lead him to develop the concept of dual nature of wave and particle. The observation of transformation of linear as well as angular momentum by radiation pressure of interacting light to the atoms [15] and other macroscopic objects play a fundamental role in fabrication of modern optomechancial devices [16]. Braginsky et al. [17, 18] performed pioneer experiments on cavity-optomechanics in which they demonstrated the dynamical effects of radiation pressure on a suspended mirror of a microwave cavity. When intra-cavity optical field interacts with the movable mirror, it triggers harmonic motion of the mirror due to the transformation of photonic momentum, as illustrated in schematic Fig. 6.3. Depending on the intra-cavity photon decay time, they demonstrate the damped and undamped motion of the moving end mirror of cavity and show that the intra-cavity radiation pressure force can be

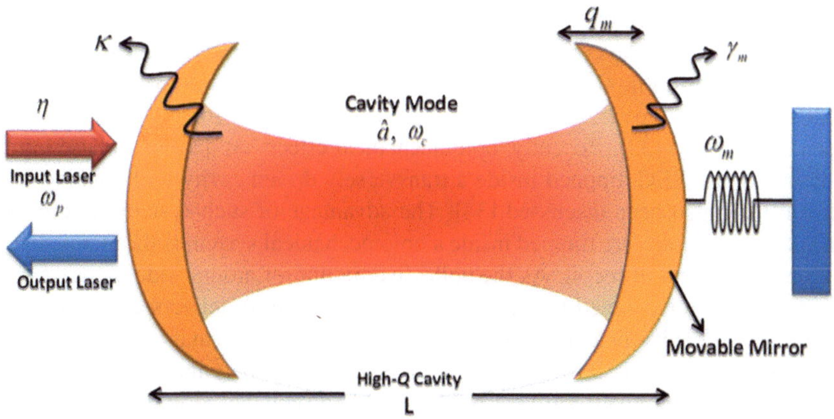

Fig. 6.3 A schematic diagram of optomechanical system consisting of a Fabry-Pérot interferometer with one movable end mirror and driven by a single mode optical field ω_p. η relates intra-cavity field with the power of external driving field and κ represents the cavity field decay rate. The cavity mode $\hat{a}$ oscillating at ω_c exerts radiation pressure on the cavity movable mirror and transfers photonic momentum which results in the oscillation of mirror with frequency ω_m and displacement q_m. γ_m is the damping rate for the mechanical motion of the mirror. Cavity quality factor can be defined by the ratio of mirror frequency and damping rate $Q = \omega_m/\gamma_m$

enhanced by using a ultrahigh-finesse cavities. These mechanical effects of light have also been demonstrated in a cavity with a kg-scaled movable mirror [19].

Further, Braginsky et al. also discussed the basic reason behind the occurrence of quantum fluctuation during the radiation force interaction and developed a condition to accurately measure the position of oscillating optomechanical mirror, especially in presence of quantum noises [20]. These investigations stood in the foundation of many recent milestones of physics, including the detection of gravitational wave by LIGO and VIRGO, where these conditions have been used to establish quantum limits for the position of movable mirrors. Further, the mechanical effects of light in cavity-optomechanics have been used to cool the vibrational mirror of the cavity to its quantum mechanical ground state by directly using laser radiations [21–23], by inducing electronic feedback [24] and by manipulating intra-cavity dynamical back-action [25, 26].

The radiation pressure induced optomechanics of cold/ultra-cold atoms and Bose-Einstein condensates (BEC) [27, 28], where the collective density excitations of atomic ensembles mimic as an atomic mirror, and the hybrid coupling of cavity field moving end mirror along with the other physics objects have further add to the significance of this field [11]. The inclusion of cold and ultra-cold atoms inside a ultra-high finesse and strongly driven optical cavity with one moving end mirror has been the subject of many extraordinary investigations and has contributed a lot in understanding of strongly correlated many-body physics. Therefore, cavity-optomechanics with BEC has been thoroughly investigated, especially regarding dynamical behavior of BEC and cavity moving end mirror in chaotic domain which eventually leads towards the pure quantum mechanical phenomenon of dynamical localization [2, 3]. The controllability of the bistable dynamics of intra-cavity photonic mode and atomic side-modes [29, 31], which basically leads to the Hamiltonian chaos, and tunable behavior of optical nonlinearities, in the form of EIT, have also been investigated [33]. Recently, optomechanical effects on quantum phase transition in a dilute BEC, trapped inside a transversely driven cavity with one movable mirror, have also been discussed [34]. The advantage of such systems where cold and ultra-cold atoms are trapped inside a optomechanical system is the atomic back-action in the form of recoil. As the collective motion of atomic side modes mimic as another mirror coupled to the cavity field, therefore, it delivers a atomic back-action to the cavity field like moving end mirror and the cavity field acts as an spring between atomic mirror and moving end mirror. Many interesting investigations regarding such system have been performed included atom induced control in cavity-optomechanics which basically leads to the back-action cooling of cavity movable mirror [35], which further get enhanced with the inclusion of spin-orbit coupled BEC [29], and three-body hybrid entanglement [36].

Cavity-optomechanics with Rydberg atoms promises many significant advantages to the physics of strongly correlated system and is the subject of a lot of ongoing scientific research. Rydberg atoms, due to their large principal quantum number and long-range dipole-dipole interactions, appeal to the further enhancements in optomechanical phenomena as discussed above. As cavity-optomechanics provides an opportunity to precisely manipulate and measure optical effects, therefore, it also, in combination with Rydberg atoms, can play a fundamental role in further elaborating quantum nonlinear optics and atom optics. Recently, coherent as well as incoherent behavior [37] and bistable dynamics [38] of Rydberg atoms inside an optomechanical cavity with moving end mirror have been discussed where authors have developed the fundamental model and presented detailed dynamics of Rydberg-cavity-optomechanics. They consider a ensemble of ^{87}Rb Rydberg atoms trapped inside a high-finesse optical cavity with one movable side mirror and driven by a single-mode pump field Ω_c and external classical probe field Ω_p, as illustrated in Fig. 6.4a. External pump laser excite atoms from ground state $|g\rangle$ (or state $5S_{1/2}$ with manifold $|F=1\rangle$) to the first excited state $|e\rangle$ (or Rydberg state $5P_{3/2}$ with manifold $|F=2\rangle$) and then classical probe field excite Rydberg atoms from excited state $|e\rangle$ to the blockaded state $|r\rangle$ (or the Rydberg state $60S_{1/2}$) which leads to the formation of a super-atom, as shown in Fig. 6.4b. The external pump field also exerts radiation pressure force $n\hbar\omega_0/L$, where n is the intra-cavity photon number, L is the cavity length and ω_0 is the cavity field frequency, on the movable mirror and transfer photonic momentum (i.e. $2\hbar k$ per photon with wavenumber k for a intra-cavity round-trip $2L/v$) which cause the oscillation of the mirror with frequency ω_m.

The motion of cavity movable mirror is coupled to the atomic mode with the help of effective cavity field ω_0 which acts as a spring between atomic mirror and cavity movable mirror. Total system Hamiltonian can be divided into two parts corresponding to the motional energies associated with each degree of freedom and their inter

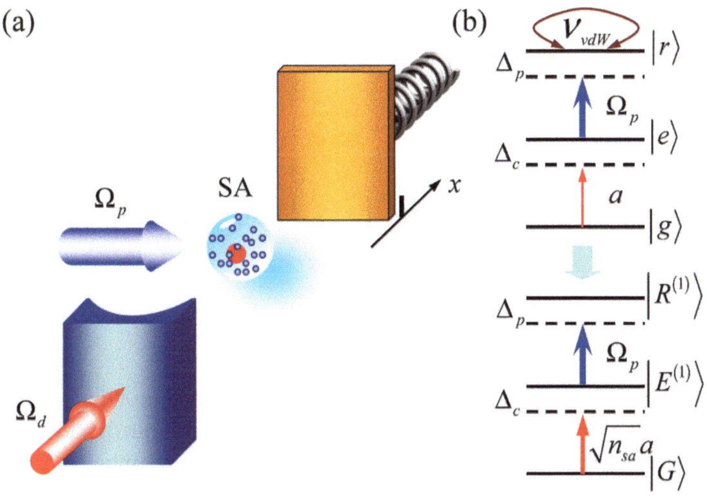

Fig. 6.4 **a** Schematic diagram of a cavity-Rydberg optomechanical system composed of a Fabry-Pérot cavity with one moving end mirror with displacement quadrature x and a Rydberg ensemble of ^{87}Rb atoms. The system is driven by a external pump field Ω_d and a classical probe field Ω_p to excite Rydberg ensemble to the blockaded state. **b** The energy level configuration indicating the transformation from three-ladder excitation (top), where the transitions $|g\rangle \rightarrow |e\rangle$ and $|e\rangle \rightarrow |r\rangle$ with detuning Δ_c and Δ_p are shown, to the Rydberg blockade excitation (bottom) under van der Waals description of dipole-dipole interactions V_{vdW}. External pump field excites atoms from ground state $|G\rangle$ to the intermediate excited state $|E^{(1)}\rangle$ with Ω_p while classical probe field cause the excitation of Rydberg blockaded state $|R^{(1)}\rangle$ from intermediate state with $\sqrt{n_{sa}}a$ under weak coupling condition between Rydberg atoms and cavity. It should be noted that this configuration only corresponds to the symmetric (or coherent) population excitation under condition of equal coupling of cavity field with all atoms in Rydberg ensemble. Adopted from [38], ©(2025) with permission from American Physical Society

and intra interactions $\hat{H} = \hat{H}_0 + \hat{H}_{int}$. By adopting JCM, the Hamiltonian $\hat{H}_0$ under rotating-wave approximation can be expressed as,

$$\hat{H}_0 = \hbar\omega_0 \hat{a}^\dagger \hat{a} + \frac{1}{2m}\left(p^2 + m^2\omega_m^2 x^2\right) + \hbar \sum_{i=1}^{N} \left(\omega_{eg}\sigma_{ee}^i + \omega_{rg}\sigma_{rr}^i\right), \tag{6.11}$$

where, as usual, $\hat{a}^\dagger$ and $\hat{a}$ are creation and annihilation operators for the cavity field defined by the commutation relation $[\hat{a}^\dagger, \hat{a}] = 1$ and $\Delta_c = \Omega_c - \omega_c$ is the detuning of cavity field with the external pump field frequency. m defines the mass while p and x are the momentum and position quadrature of the moving end mirror. $\sigma_{ee} = |e\rangle\langle e|$ and $\sigma_{rr} = |r\rangle\langle r|$ define atomic population inversion in excited state $|e\rangle$ and blockaded state $|r\rangle$ while ω_{eg} and ω_{rg} are the atomic transitional frequency from $|g\rangle \rightarrow |e\rangle$ and $|g\rangle \rightarrow |r\rangle$, respectively, where N is the total number of particles in Rydberg ensemble.

The second part of the total Hamiltonian, i.e. $\hat{H}_{int}$, can be written as,

$$\hat{H}_{int} = -\frac{\hbar\omega_0}{L}\hat{a}^\dagger\hat{a}x - \hbar\sum_{i=1}^{N}\left(\Omega_p e^{-i\Delta_p t}\sigma^i_{re} + g\hat{a}\sigma^i_{eg} + H.c.\right)$$
$$+\hbar\sum_{i<j}^{N}\frac{C_6}{R^6_{ij}}\sigma^i_{rr}\sigma^j_{rr}, \tag{6.12}$$

where first term indicates the interaction of cavity field and moving end mirror and ω_c/L defines the strength of coupling between the motion of mirror and cavity mode. The second term accommodates the interaction of Rydberg ensemble of N atoms with external probe field and pump/control field. Here it can be seen that the classical probe field trigger the transition from excited state to the blockaded state, as indicated by $\sigma_{re} = |e\rangle\langle r|$, while the external pump field cause the transition from ground state $|g\rangle$ to the excited state $|e\rangle$, shown by $\sigma_{eg} = |g\rangle\langle e|$. The parameter $g = d\sqrt{\omega_c/2\hbar V\epsilon_0}$ is the coupling strength, which can be interpreted as the optical dept of the cavity mode for atomic Rabi transitions, of atomic ensemble with intra-cavity optical mode, where d is the electric dipole moment, causing the creation of cavity-Rydberg polaritons, while V is the cavity total volume and ϵ_0 is the permittivity of the cavity vacuum state. Last term corresponds to the intra-atomic dipole-dipole interactions which are considered in van der Waals regime, discussed previously, to encounter long-range interactions, where C_6 is the van der Waals interaction coefficient and R_{ij} corresponds to the distance between two neighboring ith and jth atoms. The Rydberg-Rydberg interactions are configured in a such way, by using large values of C_6, so that only single Rydberg atoms excite to blockaded state $|r\rangle$ causing the blockage of excitation for other atoms in ensemble. It should be noted that the associated of cavity-Rydberg optomechanical system with the external pump and probe field intensities have not been considered so for simplicity but their influence can be included by using the method discussed in section of JCM [29–31, 33–36].

Further, to achieve optomechanical manipulation of such complex and strongly correlated system in a realistic domain, it is necessary to incorporate effects of damping and quantum noises associated with all mechanical degrees of freedom, especially photon shot noise and thermo-mechanical noises caused by the external heat bath. These effects can be incorporated by using Heisenberg-Langevin equation which is basically the manipulation of Langevin equation in a Heisenberg interaction picture. The Heisenberg-Langevin equations help us to reveal the stochastic time-dependent dynamical behavior of each individual degree of freedom coupled to the other subsystems,

$$\frac{\partial}{\partial t}\hat{\mathcal{O}} = \frac{i}{\hbar}[\hat{H}, \hat{\mathcal{O}}] - \Gamma\hat{\mathcal{O}} + \hat{\mathcal{N}}, \tag{6.13}$$

where $\hat{\mathcal{O}}$ is a generic operator corresponding to the any subsystem and Γ is the damping rate associated to that subsystem. $\hat{\mathcal{N}}$ accommodates the noise factor connected to the motion of each degree of freedom. By using the total system Hamiltonian $\hat{H}$, the coupled Heisenberg-Langevin equations for each component of the cavity-Rydberg optomechanical system can be written as [38],

$$\frac{\partial}{\partial t}\hat{a}(t) = -i\omega_0(1 - \frac{x(t)}{L})\hat{a}(t) + igN\sigma_{ge}(t) - \frac{\kappa\hat{a}(t)}{2} + \sqrt{\kappa}a_{in}(t), \quad (6.14)$$

$$\frac{\partial}{\partial t}x(t) = \frac{p(t)}{m}, \quad (6.15)$$

$$\frac{\partial}{\partial t}p(t) = -m\omega_m^2 x(t) + \frac{\hbar\Delta_c}{L}\hat{a}^\dagger\hat{a} - \gamma_m p(t) - \sqrt{\gamma_m}\xi(t), \quad (6.16)$$

$$\frac{\partial}{\partial t}\sigma_{ge}(t) = -(i\omega_{ge} + \gamma_e)\sigma_{ge}(t) + i\Omega_p^* e^{i\Delta_p t}\sigma_{gr}(t) + ig\hat{a}(t) + f_e(t), \quad (6.17)$$

$$\frac{\partial}{\partial t}\sigma_{gr}(t) = -(i\omega_{gr} + iS + \gamma_r)\sigma_{gr}(t) + i\Omega_p e^{-i\Delta_p t}\sigma_{ge}(t) + ig\hat{a}(t) + f_r(t). \quad (6.18)$$

The first equation reveals the time evolution of the cavity mode, where a_{in} represents Markovian noise parameter connected with the cavity field which can be defined by the delta correlation $\langle a_{in}(t)a_{in}^\dagger(t')\rangle = \delta(t - t')$ with zero mean $\langle a_{in}\rangle = 0$ under temperature condition $\hbar\omega_c >> k_B T$, where k_B is the Boltzmann constant and T is the temperature of external heat bath. κ is the cavity field decay rate which crucially depends upon the Q-factor of the cavity while γ_m is the motional damping rate of the cavity movable mirror. $\xi(t)$ is the zero-mean Langevin-force operator corresponding to the mechanical quantum noise associated with the Brownian motion of the moving end mirror, which can be express by the non-Markovian correlation function,

$$\langle\xi(t)\xi(t')\rangle = \frac{\gamma_m}{2\pi\omega_m}\int d\omega e^{-i\omega(t-t')}\left[1 + coth(\frac{\hbar\omega}{2k_B T})\right]. \quad (6.19)$$

One can note that the strength of $\xi(t)$ is directly proportional to the damping rate and inversely proportional to the frequency of the mirror which reveals that the ω should be large enough in order to suppress noise effects, under condition $\hbar\omega_c >> k_B T$. Further, γ_e and γ_r are the damping rates corresponding to the motion of Rydberg atoms in excited state $|e\rangle$ and blockaded state $|r\rangle$, respectively, caused by the external confinement of the Rydberg ensemble which are mostly adiabatically ignored in literature because of having spin independent nature [29]. Similarly, $f_e(t)$ and $f_r(t)$ are the Markovian quantum noise operators related to optomechanical motion of the Rydberg atoms in excited state $|e\rangle$ and blockaded state $|r\rangle$, respectively, defined by the delta correlation function $\langle f_{e,r}(t)f_{e,r}^\dagger(t')\rangle = \delta(t - t')$ having zero average $\langle f_{e,r}\rangle = 0$ with $\hbar\omega_{ge,gr} >> k_B T$, respectively.

The Heisenberg-Langevin equations possess a key role in order to fully understand the collective dynamics of such optomechanical system under the influences of quantum noises. The steady-state behavior of each subsystem can be easily obtained by putting time derivative equal to zero ($\partial/\partial t = 0$) in above Heisenberg-Langevin equations and these steady-state solution have been used to understand and control the stable as well as chaotic dynamics of cavity-optomechanics with ultra-cold atoms [29–31]. Further, these equations can be linearized by manipulation associated quantum fluctuations over the steady-state solutions of the system quadratures $\mathcal{O}(t) = \mathcal{O}_s + \delta\mathcal{O}(t)$, where $\mathcal{O}_s$ is the generic steady-state solution corresponding to the individual subsystem while $\delta\mathcal{O}(t)$ is the first order quantum fluctuation operator. This method has been extensively used to interpret and understate the evolution of almost all cavity-optomechanical systems with cold/ultra-cold.

6.3.1 Coherent and Incoherent Dynamics

It is interesting to discuss the coherent and incoherent dynamics of the atomic population inversion during the optomechanical interactions of cavity field with Rydberg ensemble. Such dynamics can be interrogated by analyzing the symmetric and asymmetric excitations of the atomic state. As we have already discussed that the excitation of Rydberg atoms from intermediate excited state to the Rydberg blockade states, under condition of off-resonant interaction between cavity field and atoms, cause the dissipative type of optical nonlinearities, therefore, it may leads to the excitation of asymmetric atomic state and causes incoherence in the atomic density distribution. The coherent as well as incoherent dynamics can be governed by evaluation total Hamiltonian of the system $\hat{H}$ under assumption that the all atoms are initially in ground state $|G\rangle = |g_1, g_2, ..., g_N\rangle$ and the intra-cavity optical field have same coupling strength $g = g_i$ to all atoms in Rydberg ensemble, which is experimentally possible by utilizing state-of-the-art atomic trapping methods [32]. When intra-cavity optical mode interacts with Rydberg ensemble with equal Rabi transitional force to each atom, it will excite whole ensemble to the intermediate excited state which possess symmetric population inversion, defined by the superposition principle,

$$|E^j\rangle = \frac{1}{\sqrt{N_E^j}}\left(\sum_{i=1}^{N} |e_i\rangle\langle g_i|\right)^j |G\rangle, \tag{6.20}$$

where $N_E^j = N!j!/(N-j)!$ is the normalized atomic number in that intermediate state with the total atomic number in the Rydberg ensemble [37]. After this, the external probe laser Ω_p will excite a single atom from the atomic ensemble in intermediate excited state to the Rydberg blockade, or trigger the Rydberg excitation blockade as illustrated by he white highlighted regime in Fig. 6.5a, which can similarly be fined by the symmetric superposition principle,

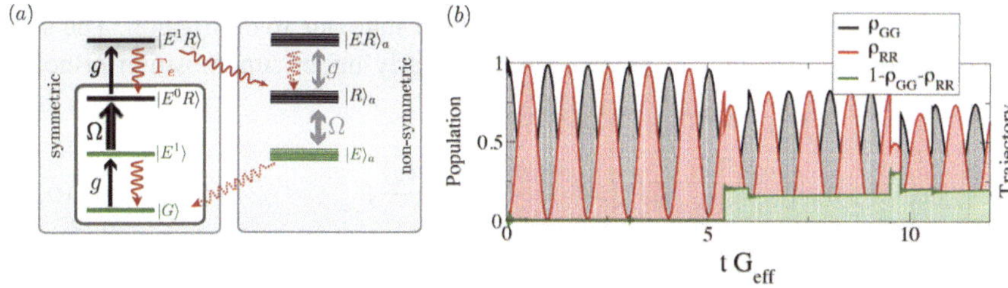

Fig. 6.5 **a** Transformation from symmetric (coherent) population excitation to the asymmetric (incoherent) population excitation of Rydberg ensemble inside an optomechanical cavity with the manipulation of radiative decay mechanism. The white inset corresponds to the symmetric ladder configuration similar to Fig. 6.4b, where the transitions $|G\rangle \rightarrow |E^1\rangle$ and $|E^1\rangle \rightarrow |E^0R\rangle$ with g and Ω are shown. The whole left side corresponds to the doubly blockade excitation to $|E_1R\rangle$ due to the radiative decay rate Γ_e, illustrated with red solid (dotted) arrows for symmetric (asymmetric) population states, which consequently results in asymmetric state excitation as shown in right side with $|E\rangle_a \rightarrow |R\rangle_a$ and $|R\rangle_a \rightarrow |ER\rangle_a$. **b** Numerical results illustrating the coherent and incoherent dynamics of Rydberg ensemble obtained from quantum Monte Carlo simulations for $N = 9$ and $\Delta_c = \Delta_e = 15\Omega$. Red and black peaks correspond to the symmetric excitation of subspaces while green region appearing at $tG_{eff} > 5$ corresponds to the asymmetric excitation of subspaces. The red and black curves are obtained by the formula $\rho_{GG} = \langle G|\rho|G\rangle$ and $\rho_{RR} = \langle E^0R|\rho|E^0R\rangle$, respectively, while green results are obtained by $1 - \rho_{GG} - \rho_{RR}$ which is supposed to be the good way to compute asymmetric excitation of subspaces. Adopted from [37], licensed under CC-BY 4.0 from IOP Publishing

$$|E^jR\rangle = \frac{1}{\sqrt{N_R^j}}\left(\sum_{i=1}^{N}|e_i\rangle\langle g_i|\right)^j\left(\sum_{i=1}^{N}|r_i\rangle\langle g_i|\right)|G\rangle, \tag{6.21}$$

where the normalized atomic number in blockaded state are defined as, $N_R^j = N \cdot N!j!/(N-j)!$. The all collective states in symmetric superposition state $|E^jR\rangle$ are excited by the only Rydberg atom, that blocks the excitation of other atoms in intermediate state, forming a Rydberg blockade and the other atoms in Rydberg ensemble either stay in the intermediate state or decay to the initial ground state $|G\rangle$. Further, as the radius of Rydberg blockade is considered larger than the size of the Rydberg ensemble itself, therefore, only one blockade excitation occurs in whole ensemble. The purely coherent dynamics can only be achievable when the only intermediate excited $|E_j\rangle$ and Rydberg $|E_jR\rangle$ superposition states are excited by the cavity-Rydberg optomechanical interactions, which is possible under condition $g = g_i$.

However, if $g \neq g_i$ and dissipation process, which is eventually caused by the doubly excited blockaded states as discussed previously, occurs then the asymmetric state may get populated which lead towards the incoherent dynamics of the Rydberg atoms in cavity-optomechanics. Figure 6.5a illustrates the occurrence of incoherent dynamics when the symmetric superposition states are transfered into asymmetric superposition states because of dissipation. Basically asymmetric states (or subspace components) are get populated by the atoms that are not permutation invariant, such as the excitation of two atoms to the intermediate state due to dissipation. The state of this second excitation can be expressed similarly under superposition principle as [37],

$$|E^1\rangle_a = \frac{|e_1 g_2\rangle - |g_1 e_2\rangle}{\sqrt{2}}, \tag{6.22}$$

which shows that the both transition have the equal probability amplitude. This process can be explained by considering a radiative decay from the intermediate excited state populated with atomic ensemble of two atoms. As the two different atoms in ensemble act differently to the radiative emission, therefore, cause the excitation of asymmetric states which finally lead to the incoherence. If we consider both atoms initially in ground state, then their interaction with cavity field transform them to the superposition state $|E^1\rangle = (|g_1 e_2\rangle + |e_1 g_2\rangle)/\sqrt{2}$, under condition that one atom will get excite to the intermediate state. After this, depending upon the cavity field, two-atom superposition state will either decay back to the ground state by emitting a photon, which have an equal probability to the decay constant Γ_e, or it get excite to the blockaded state after interacting with external probe field which can be fiend as, $|E^0 R\rangle = (|r_1 g_2\rangle + |g_1 r_2\rangle)/\sqrt{2}$, as illustrated in Fig. 6.5(a). At this point, if the two-atom ensemble get further interact with cavity field and absorb another photon then a doubly excited blockade is create in combination with intermediate state and Rydberg state, $|E^1 R\rangle = (|r_1 e_2\rangle + |e_1 r_2\rangle)/\sqrt{2}$. However, this superposition state may yield in a coupled state with $|E^0 R\rangle$ state because of the optomechanical interaction with cavity field or decay back to the ground state. In [37], authors developed another superposition state proportional to the second Rydberg state $J_e^1 |E^1 R\rangle \propto |g_1 r_2\rangle/\sqrt{2}$, with a radiative decay constant $J_e^1 = \sqrt{\Gamma_e}|g_1\rangle\langle e_1|$ which can yield in singly excited Rydberg blockaded state. The new superposition state $|g_1 r_2\rangle = (|E_0 R\rangle - |R\rangle_a)/\sqrt{2}$ can be interpreted as a combinational state of symmetric and asymmetric population excitation, having an equal probability amplitudes, which further can be reduced to $|R\rangle_a = (|r_1 g_2\rangle - |g_1 r_2\rangle)/\sqrt{2}$.

They numerically computed the behavior of coherent and incoherent dynamics of Rydberg atoms interacting with cavity-optomechanics by considering a three-level atomic system with atoms $N = 9$, as shown in Fig. 6.5b. The red and block picks correspond to the symmetric subspace trajectories while asymmetric subspaces are shown by the green region corresponding to the quantum jumps caused by the photon emission rate at atom-field coupling $tG_{eff} \sim 5$. One can clearly observe the increase incoherent dynamics due to the increase in quantum jumps caused by the asymmetric population excitation from off-resonant interaction of cavity field and Rydberg ensemble. As the population inversion from symmetric to asymmetric subspaces occurs depending upon the strength of decay rate of the intermediate state, therefore, it is possible to obtain pure symmetric transitions by suppressing the decay rate of intermediate state. In [37], they have efficiently discussed the method to slow down the decay rate of intermediate excited state with the help cavity detuning, which eventually cause the modification in atom-field coupling, and they have develop the complete set of master equation, for cavity-Rydberg optomechanics, to understand and manipulate coherent and incoherent dynamics.

6.3.2 Bistability in Rydberg-Optomechanics

In order to demonstrate and analyze optical as well as spacial bistable behavior in Rydberg-cavity-optomechanics, one should reconsider the Heisenberg-Langevin equations that describe the complete and coupled dynamics of the system. Such bistable values corresponding to the individual subsystems can be obtained by solving Heisenberg-Langevin equations in steady-state, i.e. by putting time-derivative equal to zero, which is possible by considering system quadratures as a classical variables, under mean-field approximation, and by eliminating the fast-varying components (or by inserting plane-wave ansatz instead of quantum fluctuation operators) by reassuming system quadratures as, $\hat{a} = be^{-i\omega_{cc}t}$, $\sigma_{gr} = \rho_{gr}e^{-i(\omega_{cc}+\omega_p)t}$, $\sigma_{ge} = \rho_{ge}e^{-i\omega_{cc}t}$ and $a_{in} = \alpha_{in}e^{-i\omega_{cc}t} = (\beta + \delta\beta)e^{-i\omega_{cc}t}$, where $\omega_{cc} = \omega_c(1 - \langle x\rangle/L)$ is the normalized cavity field frequency which cause the motion of cavity end mirror $\langle x\rangle \neq 0$ [38]. As the mean value of van der Waals interaction force $\langle S\rangle$ crucially depends on the number of Rydberg excitation blockades, because for a single blockade excitation that interaction force have significantly high value which makes the steady-state solution very difficult and also results in asymmetric excitations, therefore, in order to avoid this, total atomic number N in Rydberg ensemble should be small which results in small ensemble volume V less than the blockade volume V_d, having radius $r_d = (C_6|\gamma_e + i\Delta_c|/|\Omega_p|^2)^{1/6}$, blocking the excitation of other atoms in ensemble to the blockaded level. Further, to avoid the excitation of atoms to the asymmetric levels, the coupling of intra-cavity field with each atom in Rydberg ensemble should be same and atomic interaction with cavity field should be resonant to avoid dissipation. The three symmetric state of the system can then be defined as, $|G\rangle = |g\rangle^{\otimes N}$, $|E^{(1)}\rangle = (N)^{-1/2}(\sum_{j=1}^{N}|g_1, g_2, ..., e_j, ..., g_N\rangle)$ and $|R^{(1)}\rangle = (N)^{-1/2}(\sum_{j=1}^{N}|g_1, g_2, ..., r_j, ..., g_N\rangle)$.

By further redefining atomic quadratures, according to these symmetric states which forms a purely three-level ladder system under condition of weak cavity field and large atom-field detuning, as $\varrho_e = |G\rangle\langle E^{(1)}|$ and $\varrho_r = |G\rangle\langle R^{(1)}|$ instead of ρ_{ge} and ρ_{gr} respectively, the steady-state form of Heisenberg-Langevin equations can be rewritten as,

$$- m\omega_m^2\langle x\rangle + (\hbar\omega_c/L)\langle b^\dagger\rangle\langle b\rangle = 0, \tag{6.23}$$

$$ig\sqrt{N}\langle\varrho_e\rangle - \gamma_m\langle b\rangle/2 + \sqrt{\gamma_m}\beta = 0, \tag{6.24}$$

$$- (i\Delta_c + \gamma_e)\langle\varrho_e\rangle + i\Omega_p^*\langle\varrho_r\rangle + i\sqrt{N}g\langle b\rangle = 0, \tag{6.25}$$

$$- (i\Delta + i\langle S\rangle + \gamma_r)\langle\varrho_r\rangle + i\Omega_p\langle\varrho_e\rangle = 0, \tag{6.26}$$

where normalized atomic quadratures are obtained as $\langle \rho_{ge} \rangle \rightarrow \langle \varrho_e \rangle/\sqrt{N}$ and $\langle \rho_{gr} \rangle \rightarrow \langle \varrho_r \rangle/\sqrt{N}$. Here one should note that the associated quantum noises are ignored by applying classical variable approximation. $\Delta_c = \omega_{eg} - \omega_c$ is the detuning of cavity field with first pump field atomic transition while $\Delta_p = \omega_{re} - \omega_p$ is the detuning of probe field with second Rydberg transition which leads to the resultant cavity mode detuning $\Delta = \Delta_c + \Delta_p$. Interestingly, the external driving field detuning can also be defined as $\Delta_c = \Delta_0 + \omega_0 \langle x \rangle / L$, where $\Delta_0 = \omega_{eg} - \omega_0$. The total blockade population can be defined by $\varrho_{rr} = \varrho_r^\dagger \varrho_r = |\varrho_r|^2$ which possess a dual value, i.e. $\langle \varrho_{rr} \rangle = 1$ or $\langle \varrho_{rr} \rangle = 0$, depending upon the duality of dipole-dipole interaction force which can be infinite and zero, respectively. By solving steady-state Heisenberg-Langevin equations with assumption $\varrho_{rr} + \varrho_{gg} = 1$, where $\varrho_{gg} = |G\rangle\langle G|$ is the atomic population in ground state, the mean values of the system steady-state quadratures are given as [38],

$$\langle b \rangle = \frac{2\beta}{\gamma_m} + \frac{2ig\sqrt{N}}{\gamma_m} \langle \varrho_e \rangle, \tag{6.27}$$

$$\langle x \rangle = \frac{(\Delta_c - \Delta_0)L}{\omega_c} = \frac{\hbar \omega_c}{m\omega_m^2 L} \langle b^\dagger \rangle \langle b \rangle, \tag{6.28}$$

$$\langle \varrho_e \rangle = \langle \varrho_{rr} \rangle \langle \varrho_e^{2L} \rangle + (1 - \langle \varrho_{rr} \rangle) \langle \varrho_e^{3L} \rangle, \tag{6.29}$$

where terms $\langle \varrho_e^{2L} \rangle$ and $\langle \varrho_e^{3L} \rangle$ corresponding to atomic two-level and three level mean transitional operators can be expressed as,

$$\langle \varrho_e^{2L} \rangle = \frac{ig\sqrt{N}\langle b \rangle}{i\Delta_c + \gamma_e}, \tag{6.30}$$

$$\langle \varrho_e^{3L} \rangle = \frac{ig\sqrt{N}\langle b \rangle}{i\Delta_c + \gamma_e + |\Omega_p|^2 (i\Delta + \gamma_r)^{-1}}. \tag{6.31}$$

Further, by using these relations for atomic density distribution in two-level and three-level ladder configuration in atomic blockade mean steady-state solution, we obtain,

$$\langle \varrho_{rr} \rangle = \frac{Ng^2 |\langle b \rangle|^2 |\Omega_p|^2}{Ng^2 |\langle b \rangle|^2 |\Omega_p|^2 + [|\Omega_p|^2 - \Delta\Delta_c]^2 + \Delta^2 \gamma_e^2}. \tag{6.32}$$

By analyzing above equation, one can note that the dynamics of Rydberg blockade are influence by both cavity moving end mirror as well as intra-cavity optical field which eventually provides an opportunity of manipulate and utilize such complex and hybrid systems. Further, from Eq. (6.32), the susceptibility of Rydberg blockade coupled to the cavity field and moving end mirror can be easily obtained,

$$\mathcal{X} = \frac{2g\sqrt{N}}{\omega_c} \frac{\langle \varrho_e \rangle}{\langle b \rangle}, \tag{6.33}$$

where the imaginary and real part of the susceptibility $\mathcal{X}$ correspond to the absorptive and dispersive, respectively, behavior of intra-cavity optical field tune by the atom-field coupling and atomic population.

By plotting these steady-state solution as a function of effective cavity detuning Δ_0, one can analyze the occurrence of bistability in all subsystem quadratures, depending upon a particular set of characteristic parameters, as illustrated in Fig. 6.6. The black, red and blue curves correspond to the different atomic configuration including Rydberg excitation blockade case, many two-level excitation case (which basically arises from the excitation of asymmetric state by the external pump field when the cavity field coupling is different with each atom in ensemble) and many three-level excitation (which arises from the asymmetric atomic excitation with external probe field), respectively [38]. In Fig. 6.6a, one can observe the emergence of not only bistable special quadrature of cavity moving end mirror but also multiple stable solutions for moving end mirror for effective system detuning. It happens because

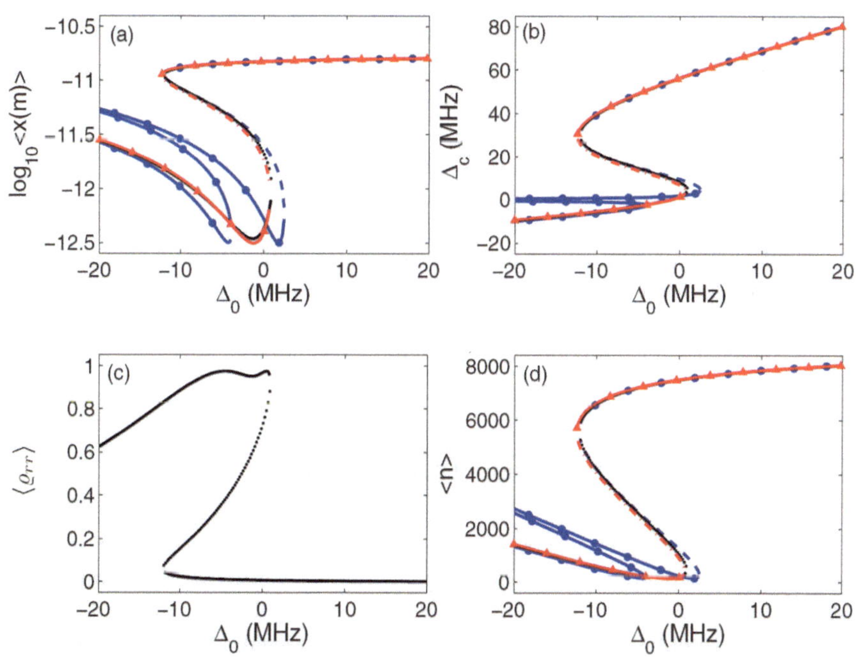

Fig. 6.6 Bistability analyses for associated subsystems of cavity-Rydberg optomechanics as a function of effective cavity detuning. **a** Cavity moving end mirror mean displacement $\langle x \rangle$ while **b–d** illustrate the external pump field detuning with cavity field Δ_c, Rydberg blockade excitation quadrature $\langle \varrho_{rr} \rangle$ and intra-cavity total photon number $\langle n \rangle$, respectively, as a function of effective system detuning Δ_0 at characteristic parameter $\beta = 1.5^4 \mathrm{Hz}^{1/2}$. Here the black, red and green curves correspond to the Rydberg blockade, multiple three-level and multiple two-level atomic configuration, respectively. The solid parts of the curves are for the stable regions while the dotted parts correspond to the unstable or chaotic regions. Adopted from [38], ©(2025) with permission from American Physical Society

when moving end mirror is only coupled to the cavity field and Rydberg blockade, or with a system of two-level configuration, then the steady-state mirror quadrature follows quadratic interaction which lead to bistability.

On the other hand, if mirror motion gets coupled to a system of three-level excitation, due to asymmetric population inversion, along with the cavity field then the mirror steady-state position faces more than quadratic interfaces which consequentially results in more than two stable states for mirror. The principle for stable and unstable state is the same discussed previously in this section which depends upon the slope and direction of the quadrature curves. Similarly, like cavity moving end mirror, the steady-state value of external pump field detuning Δ_c and intra-cavity photon number $\langle n \rangle$ follow bistable and multi-stable dynamics depending upon the atomic level configuration, as shown in Fig. 6.6b and d, respectively. It can be noted that at resonance point, i.e. $\Delta = 0$, there are two stable points for cavity mode, where the dynamics of intra-cavity photons are stable, formed by cavity field coupling with mirror and atomic degree of freedom. The lower branch of photons can be suppressed by increasing the strength of photons in upper branch by changing the atomic configuration which will in fact cause the suppression of bistable dynamics [29]. Further, Fig. 6.6c illustrates the bistable dynamics of intra-cavity Rydberg blockade in coupling with cavity mode and the moving end mirror. As Rydberg blockade excitation behaves as a single entity oscillating with cavity radiation pressure like the other two mechanical oscillators, therefore, Rydberg blockade also possess quadratic steady-state solution due to the coupling with two other mechanical oscillators. In [38], authors also investigated the influence of total atomic number on the bistable dynamics of subsystem quadratures and studied the nonlinearities by obtaining EIT from effective susceptibilities. Although, these investigations provide the basic foundation of cavity-Rydberg optomechanics but still it should be further explored specially in regard to modern quantum optics and quantum information.

6.4 Cavity-Rydberg Hybrid Transduction

Cavity-Rydberg hybrid transduction represents a rapidly emerging quantum interface that leverages the extreme sensitivity of Rydberg atoms to microwave and millimetre-wave radiation, combined with the high-finesse control offered by optical cavities, to enable coherent interconversion between vastly different frequency regimes. Rydberg states, owing to their exaggerated dipole moments and strong polarizability, can strongly couple to microwave photons, while optical cavities facilitate interaction with optical photons through cavity quantum electrodynamics (QED). This dual accessibility enables a promising transduction mechanism between optical and microwave domains—a long-sought capability for linking superconducting quantum processors with optical quantum networks.

In recent pioneering work, neutral atoms prepared in Rydberg states were embedded in an optical cavity and exposed to millimetre-wave radiation, demonstrating the first **quantum-enabled transduction** between millimetre-wave and optical fields [39]. This platform operated by inducing transitions between Rydberg levels using microwave fields, while simultaneously coupling optical transitions to a high-finesse optical cavity. Quantum coherence was preserved during the transduction process, as shown by the measurement of single-photon-level signals and preservation of quantum statistics.

Other platforms have explored the integration of superconducting atom chips with Rydberg atoms to demonstrate **cavity-driven Rabi oscillations** between Rydberg states via microwave resonators [40]. Such systems enable deterministic control over atom—photon interactions and have the potential to be integrated with optomechanical or fiber-coupled systems. Additionally, optical bow-tie cavities with ultra-high finesse have been developed to couple single Rydberg atoms to cavity modes with high cooperativity, achieving near-unit photon emission probability and enabling strong light—matter coupling even at the single-atom level [41]. These advancements set the foundation for scalable, room-temperature, quantum transduction systems that combine the spectral flexibility of Rydberg atoms with the coherent processing capabilities of optical cavities.

Overall, cavity-Rydberg hybrid transduction stands at the confluence of quantum optics, atomic physics, and quantum information science, offering a route toward broadband and noise-resilient interconnects across quantum platforms. Its development is expected to play a central role in quantum network architectures and hybrid quantum technologies.

6.4.1 *Hybrid Millimetre Wave to Optical Transduction*

In this section, we provide a detailed description and analysis of the experimental system and results presented by Kumar et al. [39], where a quantum transducer bridges the millimetre wave (mm-wave) and optical domains by exploiting strong coupling of neutral Rydberg atoms to dual electromagnetic cavities. This transduction mechanism is a critical advancement for linking microwave quantum processors to optical quantum communication channels. The transduction platform combines a cloud of ultracold ^{85}Rb atoms positioned at the intersection of two distinct cavities: a superconducting millimetre wave resonator and a high-finesse optical cavity. Figure 6.7 provides a comprehensive overview of the apparatus.

The Fig. 6.7a depicts a schematic cross-sectional view of the hybrid system. The mm-wave cavity is a three-dimensional superconducting resonator operating around 50 GHz, housed inside a dilution refrigerator to minimize thermal photons. The optical Fabry-Pérot cavity, orthogonal to the mm-wave cavity, supports a fundamental mode resonant near 780 nm. The ultracold atomic ensemble is trapped at the spatial overlap of these two cavity modes, allowing simultaneous coupling to both electromagnetic fields. On other hand, Fig. 6.7b illustrates the atomic energy level diagram central to the transduction process is illustrated here. The atoms are prepared in the ground state $|g\rangle$ and excited to the intermediate state $|e\rangle$ via a classical control laser with Rabi frequency Ω_c. From $|e\rangle$, atoms are excited to a highly-excited Rydberg

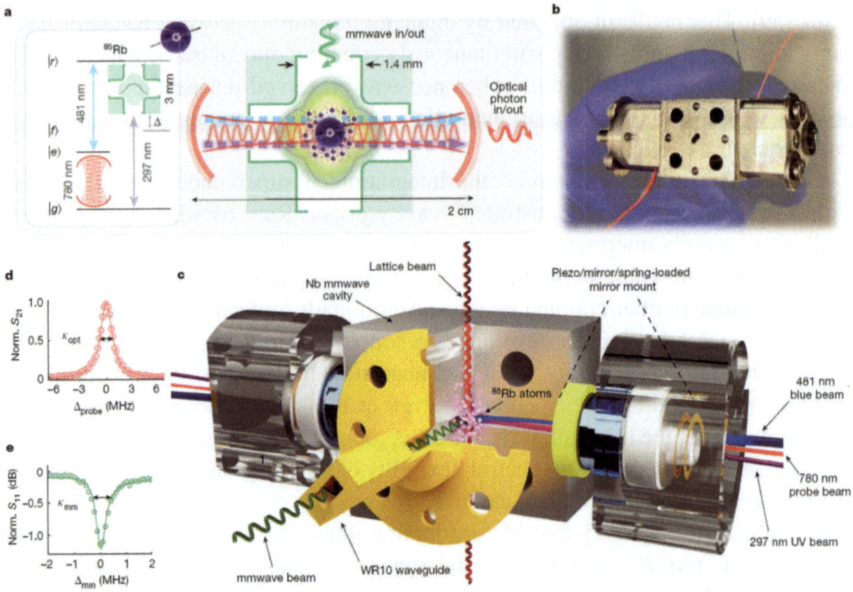

Fig. 6.7 Experimental setup and atomic structure of the quantum transducer. **a** Dual cavity geometry with ultracold ^{85}Rb atoms at the mode intersection. **b** Four-level ladder atomic scheme: ground $|g\rangle$, excited $|e\rangle$, Rydberg $|r\rangle$ and second Rydberg $|r'\rangle$ states coupled by optical and mm-wave fields. **c** Mode profiles of mm-wave (blue) and optical (red) cavities overlapping the atomic cloud. **d** Cryogenic superconducting mm-wave cavity assembly. Adopted from [39], ©(2025) with permission from Springer Nature

state $|r\rangle$, which strongly couples to another nearby Rydberg state $|r'\rangle$ through the mm-wave cavity mode at frequency ω_μ. This ladder-type configuration facilitates coherent absorption of mm-wave photons and subsequent optical emission.

The Fig. 6.7c visualizes the spatial mode profiles of both the mm-wave and optical cavity fields, demonstrating their optimized overlap with the atomic cloud. The efficient mode matching is crucial to maximize the collective atom-photon coupling strengths g_μ and g_o, which scale as $g_i \propto \sqrt{N}$ with atom number N. While Fig. 6.7d provides details of the superconducting resonator are shown, including the cryogenic environment that cools the cavity to milliKelvin temperatures, reducing thermal photon noise and improving coherence times. The mm-wave cavity has a high quality factor $Q_\mu \sim 10^5$, critical for strong light-matter interaction.

The interaction Hamiltonian governing the system in the rotating frame reads:

$$\hat{H}_{\text{int}} = \hbar g_o \left(\hat{a}^\dagger \hat{S}_{ge} + \hat{a}\hat{S}_{eg}\right) + \hbar g_\mu \left(\hat{b}^\dagger \hat{S}_{rr'} + \hat{b}\hat{S}_{r'r}\right) + \hbar \Omega_c \left(\hat{S}_{er} + \hat{S}_{re}\right), \quad (6.34)$$

where $\hat{a}$ ($\hat{a}^\dagger$) and $\hat{b}$ ($\hat{b}^\dagger$) are the annihilation (creation) operators of the optical and mm-wave cavity modes, respectively; g_o and g_μ are the single-atom coupling strengths; and $\hat{S}_{ij}$ are collective lowering/raising operators between atomic states $|i\rangle$ and $|j\rangle$.

The dynamics are captured by coupled quantum Langevin equations, incorporating cavity loss rates κ_o and κ_μ and atomic decoherence γ_i. The system operates in the strong-coupling regime, enabling coherent energy exchange between the microwave and optical fields mediated by the Rydberg ensemble.

The experimental results demonstrating the transduction performance are summarized in Fig. 6.8. The Fig. 6.8a shows the quantum conversion efficiency $\eta(\Delta)$ as a function of detuning $\Delta = \omega_\mu - \omega_{rr'}$ between the mm-wave cavity mode frequency and the Rydberg-Rydberg transition frequency. The peak efficiency of approximately 58% occurs at near-zero detuning ($\Delta \approx 0$), indicating near-resonant energy exchange. The Lorentzian shape of the efficiency curve reflects the combined linewidths of the mm-wave cavity and the Rydberg transition. The solid curve represents a theoretical fit based on a master equation model including all decay and decoherence channels, demonstrating excellent quantitative agreement with experimental data points. Efficiency diminishes away from resonance due to reduced coupling and increased off-resonant losses.

The Fig. 6.8b quantifies the noise added by the transducer during conversion, expressed in equivalent photons n_{add}. Remarkably, n_{add} remains below one quantum over the bandwidth of operation, approaching the quantum limit for ideal transduction. Noise contributions arise from thermal photons in the mm-wave cavity, atomic spontaneous emission, and technical laser noise. The experimental data (dots) align well with theory (solid line), confirming that the cryogenic environment and atomic coherence effectively suppress noise. The low added noise underscores the platform's suitability for quantum information transfer.

The Fig. 6.8c demonstrates preservation of the quantum statistical properties of photons during transduction by measuring the normalized second-order correlation function of the optical output. The observed antibunching dip at zero delay $g^{(2)}(0) < 1$ confirms that single-photon character is maintained after conversion. The shape and width of $g^{(2)}(\tau)$ reflect the coherence time of the converted photons, limited by cavity linewidths and atomic decoherence. This measurement conclusively proves that the transducer not only transfers energy but also preserves quantum coherence and nonclassical photon statistics. The inset compares the $g^{(2)}(\tau)$ of input mm-wave photons and output optical photons, highlighting the fidelity of quantum state transfer.

The high conversion efficiency coupled with low added noise demonstrates the coherent nature of the process, enabled by the large electric dipole moments of Rydberg states and strong atom-cavity coupling. The multimode cavity architecture provides spectral and spatial mode matching essential for efficient energy exchange. The ladder-type atomic configuration, combined with the control laser, realizes an effective three-wave mixing process:

$$\text{mm-wave photon} + \text{control laser photon} \leftrightarrow \text{optical photon},$$

facilitated by the atomic coherence between $|g\rangle$, $|e\rangle$, and $|r\rangle$ states. The transduction bandwidth and efficiency are tunable by adjusting the control laser power and cavity detunings. The experiment represents a crucial step toward building hybrid quantum

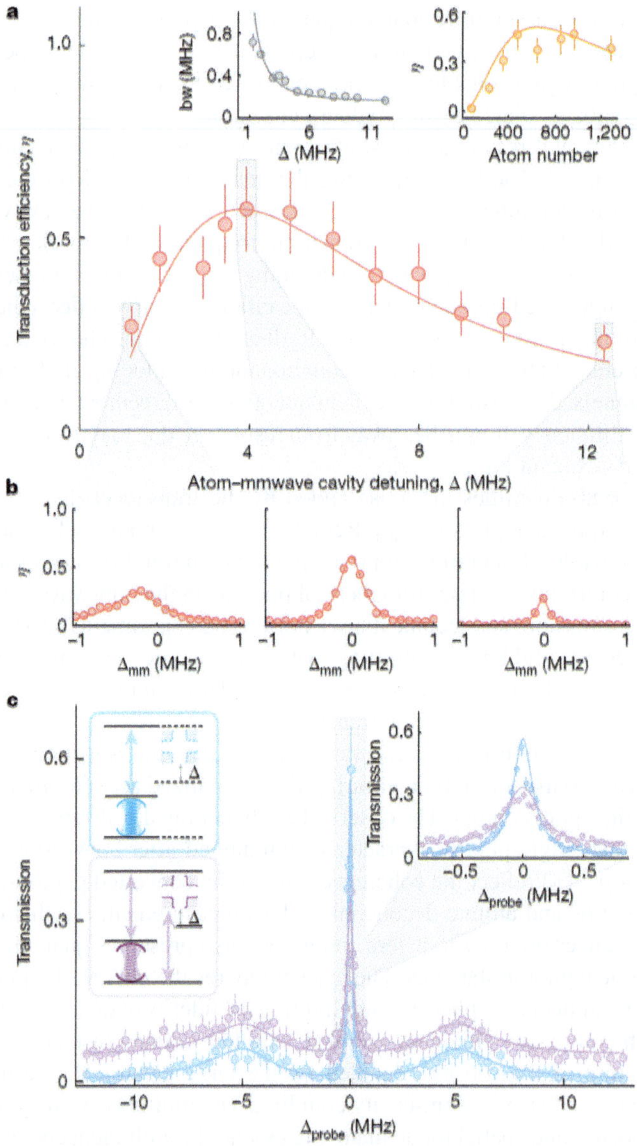

Fig. 6.8 Key performance metrics of the transducer. **a** Conversion efficiency η versus mm-wave detuning Δ, showing peak near resonance. **b** Added noise photon number n_{add} as a function of detuning, approaching the quantum limit. **c** Second-order correlation function $g^{(2)}(\tau)$ of converted optical photons, demonstrating preservation of quantum coherence and antibunching. Insets compare input mm-wave and output optical photon statistics. Obtained from [39], ©(2025) with permission from Springer Nature

networks by providing an interface between superconducting qubits operating in the microwave domain and optical photons suitable for long-distance transmission through fiber optics.

The work by Kumar et al. [39] represents a significant milestone in hybrid quantum technologies by demonstrating an efficient, low-noise quantum transducer between millimetre wave and optical frequencies using neutral Rydberg atoms. Through the clever integration of ultracold atomic ensembles with high-quality superconducting and optical cavities, the system harnesses the strong dipole interactions and long coherence times of Rydberg states to achieve coherent energy conversion with efficiencies exceeding 50% and added noise below one quantum. The preservation of nonclassical photon statistics during transduction highlights the device's capability to faithfully transfer quantum information across disparate frequency regimes. This approach paves the way toward scalable quantum networks that interface microwave quantum processors with optical communication channels, overcoming a critical bottleneck in quantum information science. Future improvements in cavity design, atomic control, and cryogenic engineering will likely enhance bandwidth and efficiency further, making such Rydberg-based transducers essential components in next-generation quantum technologies.

References

1. M. El Ghafar et al., Dynamical localization in the Paul trap. Phys. Rev. Lett. **78**, 4181 (1997)
2. K.A. Yasir, M. Ayub, F. Saif, Exponential localization of moving end mirror in optomechanics. J. Mod. Opt. **61**, 1318 (2014)
3. M. Ayub, K.A. Yasir, F. Saif, Dynamical localization of matter waves in optomechanics. Laser Phys. **24**, 115503 (2014)
4. H.M. Gibbs, S.L. McCall, T.N.C. Venkatesan, Differential gain and bistability using a sodium-filled Fabry-Perot interferometer. Phys. Rev. Lett. **36**, 1135 (1976)
5. W. Lange et al., Optical bistability with Rydberg atoms. J. Phys. Colloques **49**(C2), 81–84 (1988)
6. F. Casagrande, L.A. Lugiato, W. Lange, H. Walther, Multistability in mesoscopic Rydberg-atom systems. Phys. Rev. A **48**, 790 (1993)
7. C. Balconi et al., Self-pulsing behaviour in a mesomaser. Opt. Commun. **114**, 425–431 (1995)
8. M. Benassi, F. Casagrande, W. Lange, Atomic cooperation and nonlinear dynamics in a mesoscopic maser. Quantum Semiclass. Opt. **9**, 879–898 (1997)
9. S.S. Hassan, Y.A. Sharaby, Mesoscopic multistability with atomic coherence effects. J. Opt. B: Quantum Semiclass. Opt. **7**, S682–S688 (2005)
10. M. Aspelmeyer, T.J. Kippenberg, F. Marquardt, Cavity optomechanics. Rev. Mod. Phys. **86**, 1391 (2014)
11. P. Meystre, A short walk through quantum optomechanics. Ann. Phys. **525**, 215 (2013)
12. J. Kepler, De cometis libelli tres, Typis Andreae Apergiri (1619)
13. P. Lebedew, Experimental examination of light pressure. Ann. Phys. **311**, 433 (1901)
14. A. Einstein, On the development of our understanding of the nature and composition of radiation. Z. Phys. **10**, 817 (1909)
15. R. Frisch, Experimenteller Nachweis des Einsteinschen Strahlungs*rückstoβ*es. Z. Phys. B **86**, 42 (1933)
16. R.A. Beth, Mechanical detection and measurement of the angular momentum of light. Phys. Rev. **50**, 115 (1936)

17. V.B. Braginskii, A.B. Manukin, Ponderomotive effects of electromagnetic radiation. JETP **25**, 653 (1967)
18. V.B. Braginskii, A.B. Manukin, M.Y. Tikhonov, Investigation of dissipative ponderomotive effects of electromagnetic radiation. JETP **31**, 829 (1970)
19. B.D. Cuthbertson, M.E. Tobar, E.N. Ivanov, D.G. Blair, Parametric back-action effects in a high-Q cyrogenic sapphire transducer. Rev. Sci. Instrum. **67**, 2435 (1997)
20. V.B. Braginsky, F.. Ya.. Khalili, Quantum nondemolition measurements: the route from toys to tools. Rev. Mod. Phys. **68**, 1 (1996)
21. A.D. O'Connell et al., Quantum ground state and single-phonon control of a mechanical resonator. Nature **464**, 697 (2010)
22. J.D. Teufel et al., Sideband cooling of micromechanical motion to the quantum ground state. Nature **475**, 359 (2011)
23. J. Chan et al., Laser cooling of a nanomechanical oscillator into its quantum ground state. Nature **478**, 89 (2011)
24. C. Sch$\ddot{a}$fermeier, et al., Quantum enhanced feedback cooling of a mechanical oscillator using nonclassical light. Nat. Commun. **7**, 13628 (2016)
25. T.J. Kippenberg, K.J. Vahala, Cavity optomechanics: back-action at the mesoscale. Science **321**, 1172 (2008)
26. A. Schliesser et al., Radiation pressure cooling of a micromechanical oscillator using dynamical backaction. Phys. Rev. Lett. **97**, 243905 (2006)
27. K.W. Murch et al., Observation of quantum-measurement backaction with an ultracold atomic gas. Nat. Phys. **4**, 561 (2008)
28. F. Brennecke, S. Ritter, T. Donner, T. Esslinger, Cavity optomechanics with a Bose-Einstein condensate. Science **322**, 235 (2008)
29. K.A. Yasir, W.M. Liu, Tunable bistability in hybrid Bose-Einstein condensate optomechanics. Sci. Rep. **5**, 10612 (2015)
30. K.A. Yasir, Opt. Commun. **488**, 126820 (2021)
31. K. Zhang, W. Chen, M. Bhattacharya, P. Meystre, Hamiltonian chaos in a coupled BEC-optomechanical-cavity system. Phys. Rev. A **81**, 013802 (2010)
32. T.P. Purdy et al., Tunable cavity optomechanics with ultracold atoms. Phys. Rev. Lett. **105**, 133602 (2010)
33. K.A. Yasir, W.M. Liu, Controlled electromagnetically induced transparency and Fano resonances in hybrid BEC-optomechanics. Sci. Rep. **6**, 22651 (2016)
34. Z. Wang et al., Collapse of the superradiant phase and multiple quantum phase transitions for Bose-Einstein condensates in an optomechanical cavity. Phys. Rev. A **93**, 033630 (2016)
35. M. Paternostro, G.D. Chiara, G.M. Palma, Cold-atom-induced control of an optomechanical device. Phys. Rev. Lett. **104**, 243602 (2010)
36. M. Asjad, F. Saif, Steady-state entanglement of a Bose-Einstein condensate and a nanomechanical resonator. Phys. Rev. A **84**, 033606 (2011)
37. A. Carmele, B. Vogell, K. Stannigel, P. Zoller, Opto-nanomechanics strongly coupled to a Rydberg superatom: coherent versus incoherent dynamics. New J. Phys. **16**, 063042 (2014)
38. D. Yan et al., Duality and bistability in an optomechanical cavity coupled to a Rydberg superatom. Phys. Rev. A **91**, 023813 (2015)
39. A. Kumar et al., Quantum-enabled millimetre wave to optical transduction using neutral atoms. Nature **615**, 614–619 (2023)
40. M. Kaiser et al., Cavity-driven Rabi oscillations between Rydberg states of atoms trapped on a superconducting atom chip. Phys. Rev. Res. **4**, 013207 (2022)
41. Y.T. Chen et al., High-finesse bow-tie cavity for strong atom-photon coupling in Rydberg arrays, arXiv:2207.06876 [quant-ph] (2022)

Chapter 7
Cavity-Rydberg Quantum Information and Technology

Abstract The final chapter addresses the application of cavity-Rydberg physics to quantum information science. It explores coherent phenomena ranging from Rabi oscillations to the generation of Schrdinger cat states. Various entanglement protocols, including photon-induced and collision-induced Rydberg entanglement, are developed alongside proposals for quantum phase gates, coherent control, and single-shot optical π phase shifts. Schemes for scalable architectures, such as quantum-network registers using optical tweezers and error-detected quantum operations, are presented. The chapter concludes with approaches to quantum tomography of multi-atom systems, underlining the role of cavity-Rydberg platforms as building blocks for scalable quantum technologies.

7.1 Introduction

The quantum information process or science fundamentally deals with the utilization of quantum principle of physics to transfer and compute information by manipulating the quantum superposition concept of atomic states and the entangled characteristics of matter and light [1]. The information is converted into quantum basic units, i.e. quantum bits or qubits, which are basically two-level quantum systems or combined states of atoms and photons in a two-level in a unique configuration, like Rydberg polaritons. As these qubits are teleported from one place to another through quantum entanglement, therefore, quantum information promises a fast and secure communication and computation which significantly appeals to the modern information technologies. In quantum domain, multiple qubits are manipulated in superposition form having a certain probability amplitude for each qubit, which can be intuitively used as quantum logic gates where the output depends on the probability amplitudes [2]. As, unlike classical logic gates, the logic operations of quantum gates are performed by the entanglement between qubits [3, 4], which makes them very complex in nature and allows to perform reversible computation and provides the opportunity of multiple choice in output. For example, if two or more than two qubits are entangled with each other, then one can perform logic decisions by changing the state of one

K. A. Yasir and W.-M. Liu, *Rydberg Atoms in Cavity*, Springer Series in Solid-State Sciences 213, https://doi.org/10.1007/978-981-95-4571-1_7

qubit which will constituently results in similar change in all other entangled qubits. Such entangled manipulation of qubits and the states of matter have make possible to test and verify quantum cryptography [5, 6], quantum distant state teleportation for communication [7] and many other modern quantum phenomena [8, 9].

The major hurdle that limits the practical implementation of quantum information and computation processes is the phenomenon of decoherence which destroys the quantum control to manipulate this phenomenon and causes the destruction of entanglement between associated degrees of freedom [10, 11]. Although some of the techniques have been used to conditionally suppress decoherence to some extent to demonstrate some of the quantum information processes, like quantum cryptography [12] and teleportation [13, 14], the cavity quantum electrodynamics provides a perfect and general tool to handle the phenomenon of decoherence [15, 16]. The cavity quantum electrodynamics in presence of coupled atomic beam offers a better platform to understand matter-field interactions and manipulates these interactions to coherently perform phenomenon of quantum information. In such systems, a single atom or a single photon can be interpreted as a qubit strongly coupled to the cavity field or atomic motion, as explained by the JCM in Sect. 3.2. Especially, in the case of Rydberg atoms due to their high principal quantum number and blockade characteristics, these quantum bits can be implemented at the size of blockade and can excite a millimeter wave single photon, where the whole atomic ensemble acts as an effective two-level system, as previously discussed in Sects. 2.4 and 4.3. As Rydberg atoms together with cavity quantum electrodynamics demonstrate dispersive and dissipative types of quantum nonlinear optics, see Chap. 4, therefore, the phenomenon of quantum information can be studied and tested in strongly interacting cavity-Rydberg setups. Rydberg atoms inside the cavity also enhance the control and manipulation of cavity photon Fock state, see Sect. 3.4. Further, the experimental demonstration of coherent generated cavity-Rydberg masers, see Sect. 3.5, also amplifies the significance of cavity-Rydberg systems in quantum information.

Cavity quantum electrodynamics with Rydberg atoms has been thoroughly investigated initially by Haroche et al. where they tested the fundamental concepts of quantum entanglement, including atom-atom [17, 18], atom-field [19–21] and many-body entanglement [22]. They first time tested experimentally the quantum phase gates [23] by manipulating quantum coherence and decoherence with cavity-Rydberg system. In this part of the book, we will briefly review the advancements in cavity-Rydberg quantum information. As the purpose of this book is to discuss general and overall aspects of Rydberg atoms in cavities. As the applications of Rydberg atoms have already been extensively discussed [24], therefore, we will focus on the major and necessary concepts of quantum information in order to provide readers a vast understanding of cavity-Rydberg quantum electrodynamics.

7.2 Rabi Oscillations to Schrödinger Cats

The concept of quantum superposition is fundamentally based on the coexistence of an output (or physical object) in two, or more than two, quantum states having a certain probability of finding that output or object in each quantum state. That superposition state of multiple quantum states are bounding by the observation of measurement which means if we perform measurement, that superposition state will collapse and we will find quantum particle in one of the quantum states. The concept of quantum superposition was demonstrated by Erwin Schrödinger in 1935 in his thought experiment where he visualize superposition principle by putting an alive cat in closed box along with a flask of poison, a radioactive source and radioactive detector, in such position that if detector detects radiation (or single atom decay occurs) then it will break the flask of poison which will eventually kill the cat. Now the life of the cat crucially depends upon that radioactive source which is highly unpredictable because the decay of an atom highly dependent on the quantum interaction or Rabi transitions. Therefore, the cat will exists in both states, i.e. alive or dead, at the same time having equal probability, defined by that Rabi transitions, until we open the box and measure that whether the cat is dead or not and at that point superposition state of the cat will collapse. This thought experiment is referred to Schrödinger cat or Schrödinger cat paradox and significantly explains the phenomenon of quantum superposition that occurs in atomic (or photonic) states and stands in the foundations of quantum mechanics and quantum information. The concept of Schrödinger cats arises from Einstein-Podolsky-Rosen article (EPR paradox) [25] in which Einstein and coauthors debate on the wave function theory, based on the quantum superposition principles and entanglement, and illustrate that the wave function theory is either incomplete or there are some hidden variables corresponding to the reality.

The quantum superposition plays an important role in defining and evaluating quantum bits in quantum information. For example, if we consider qubits $|0\rangle$ and $|1\rangle$, like classical bits, then according to the quantum superposition theorem information (or outcome) can be in combination of these both states $(|0\rangle + |1\rangle)\sqrt{2}$ having equal probability amplitudes, i.e. $1/\sqrt{2}$. On the other hand, in atomic system (or subsystem), such kind of Schrödinger cats can be prepared by considering atomic either in ground state $|g\rangle$ or in excited state $|e\rangle$. Similarly, in cavity-Rydberg system, the superposition state of Rydberg atoms (or blockade) can be constructed in combination with cavity field, as discussed in Sect. 4.4.1, and was first experimentally demonstrated by Haroche et al. where they considered a cavity-Rydberg system composed of two superconducting mirrors oriented in a Fabry-Pérot configuration and a beam of Rydberg atoms, having circular configuration, passes through the cavity, as illustrated in Fig. 1.1. In their experiments, the intra-cavity fields can have the wavelength up to 6 mm, oscillating at frequency 51.1 GHz, and interacts with the single Rydberg atoms, having principal quantum number up to 51–50, at the center of the cavity.

Further, they cool the system to about $\sim$1 K to avoid thermal noises due to blackbody radiations. When intra-cavity optical mode coherently and resonantly interacts with Rydberg atom(s), it will trigger the atomic transition from ground state $|g\rangle$ to the Rydberg excited state $|e\rangle$, as define by the JC interaction Hamiltonian 3.2,

$$\hat{H}_{int} = -\frac{\hbar\Omega}{2}\left[\hat{a}^{\dagger}|e\rangle\langle g| + \hat{a}|g\rangle\langle e|\right], \tag{7.1}$$

where $\hat{a}^{\dagger}$ and $\hat{a}$ are photonic creation and annihilation operators causing the transition of an atom from excited to ground and ground to excited, respectively. Ω is the vacuum Rabi transitional frequency which determines the coupling of atomic mode with the intra-cavity field. When Rydberg atom interacts with cavity field, it initially emits a photon to the empty cavity and then reabsorbs the same photon and cause the quantum Rabi oscillations of atomic mode between ground state $|g\rangle$ and excited state $|e\rangle$.

The quantum Rabi oscillation are the basic ingredient to prepare atom or photon qubits and engineer entanglement between atom-field and atom-atom [26, 27], when two ground state atoms interacts with cavity field, as illustrated by Schrödinger cats paradox (also see Sect. 3.3.1). By changing the duration of intra-cavity field and Rydberg atom interaction and measuring the ionization signal from the cavity, one can determine the behavior of quantum Rabi oscillation as a function of time, as illustrated in Fig. 7.1a. These Rabi oscillation of atomic transition can be altered by simply exerting electromagnetic field to the cavity mirror that results in Stark scattering and cause the Stark shift in cavity field, as discussed previously, providing damping to the Rabi oscillations [27], as illustrated in Fig. 7.1b–d, where the sinusoidal nature of Rabi oscillation has been destroyed with the injection of external electromagnetic signal.

These Rabi oscillations are enough to interpret superposition behavior of Rydberg atoms in cavity which can be extended to the quantum gate formation. For example, after $\pi/2$ cycle of the Rabi oscillation, as mentioned by a red dot in Fig. 7.1a, one can construct the ideal coherent combined state of Rydberg atom and cavity field,

$$|C\rangle = \frac{1}{\sqrt{2}}\left[|e, 0\rangle + |g, 1\rangle\right], \tag{7.2}$$

where the excited state $|e\rangle$ is correlated to the intra-cavity vacuum state, i.e. when Rydberg absorbs cavity photon and excites to $|e\rangle$, and ground state $|g\rangle$ is correlated to a cavity photon, which Rydberg emits while moving to the ground state. The superposition of these coherent states can be referred to a entangled Schrödinger states of cavity field and an atom which may survives when the atoms leaves the cavity. After the π Rabi cycle, the coupling strength of atomic and intra-cavity photonic coherent states can be exchanged which results in the phenomenon of entanglement swapping. Further, if consider the quantum Rabi oscillation after 2π cycles then it is possible to change the notation of these coherent cat states which consequently gives us the liberty to rotate the spin of quantum particles (or subsystems). These descrip-

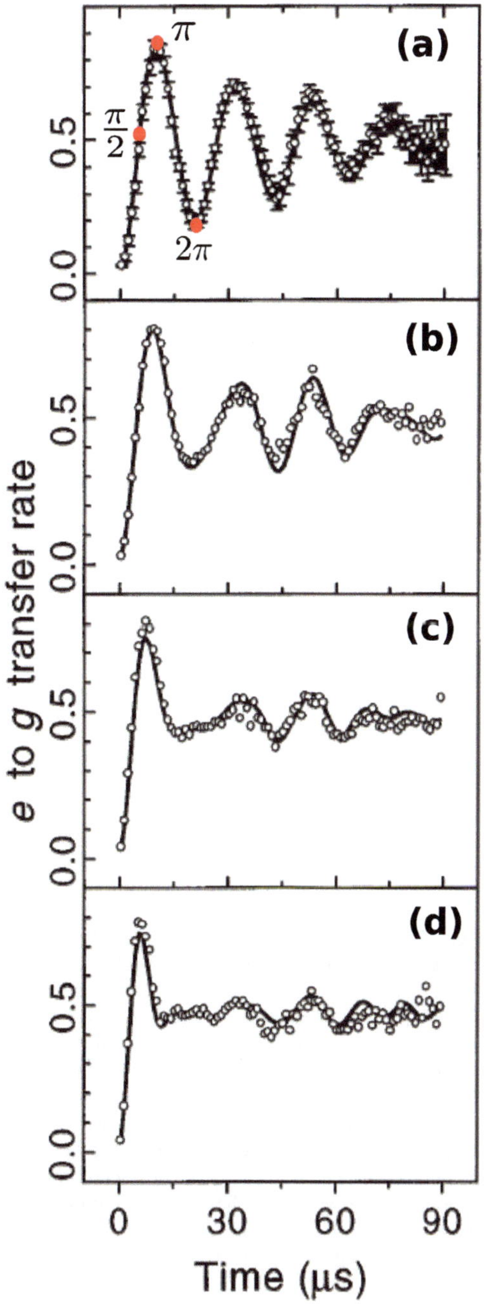

Fig. 7.1 The quantum Rabi oscillation of a Rydberg atoms between ground state $|g\rangle$ and excited state $|e\rangle$ induced by the intra-cavity optical field. **a** The absence of the external electromagnetic field to the cavity mirrors which only leaves thermal excitation to cause quantum Rabi oscillation with an average thermal photon number 0.06(±0.01). The red dots correspond to the $\pi/2$, π and 2π Rabi cycles of the oscillations as mentioned in plot. **b–d** The intra-cavity atomic Rabi oscillation in presence of external electromagnetic field 0.40(±0.02), 0.85(±0.04) and 1.77(±0.15), respectively, which cause the Stark effects in cavity field and damps the Rabi oscillations. The dotted curves correspond to the experimental results while the solid curves represent theoretical values. Adopted from [27], ©(2025) with permission from American Physical Society

tions provides us all tools that are required to perform quantum logic operations in quantum gates with the help of atomic ($|g\rangle$ and $|e\rangle$) and photonic ($|0\rangle$ and $|1\rangle$) qubits.

By following above discussed mechanism, Haroche et al. [28] experimentally demonstrated the coherent Schrődinger cat states and intra-cavity photon Fock states [29, 30], as explained in Sect. 3.4, in a cavity-Rydberg system and illustrates the dependence of these coherent states on different phases of intra-cavity atom-field interactions. They prepare circular Rydberg atoms in the sub-cavities, as illustrated by R_1 and R_2 in Fig. 7.2a and send them to the main cavity C, where they coherently interact with the cavity field, and detect the ionization with detector D. They computed the dynamics of these states by observing the evolution of intra-cavity atomic

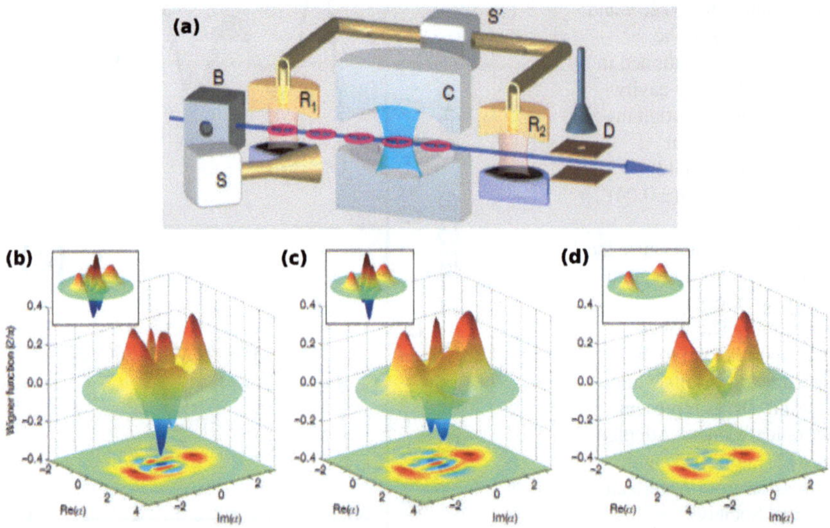

Fig. 7.2 The schematic diagram of cavity-Rydberg system and Wigner function for the cavity-Rydberg Schrődinger cat states. **a** The schematic configuration where coherent beam of Rydberg atoms coming from atomic source B is prepared in circular configuration in sub-cavities R_1 and R_3 with the help of electromagnetic source S' and then send to the main cavity C. The light from source S is shined on one of the cavity C mirrors that after diffracting from the mirror drives the cavity providing a coherent control over the phase and amplitude. The source S is used to excited the cavity and reconstruct the Schrődinger cat states of Rydberg atoms inside cavity. Finally, the output is measured by the detector D. **b** and **c** the Wigner function evolution as a function of real and imaginary values of α to reconstruct the even (when the number of photon is even) and odd (when the number of photon in cavity is odd) Schrődinger cat states of Rydberg atoms, respectively, at $n_m = 3.5$ and $\mathcal{X} = 0.37\pi$. The effective detuning $\delta/2\pi = 51$kHz of the system and interferometer phase $\phi = -\Phi(0, \phi) + \pi$ are kept constant during the reconstruction process. The experimental results are averaged over 400, where the total number of atoms observed in each sample are 2, 000 and the dimensional order of Hilbert space is taken as 11. **d** The Wigner description of the intra-cavity field in the form of cat states when Rydberg atom leaves the cavity without disturbing the effects of ground and excited states. The color configuration corresponds to the strength of the Wigner function and the inset figures illustrate theoretical calculation as a comparison. Adopted from [28], ©(2025) with permission from Springer Nature

density matrix or the Wigner function. The Wigner function of the semi-classical states takes a Gaussian shape while the Wigner description of photonic Fock states and Schrödinger cat states contains oscillation and negative values corresponding to the quantum interferences, as can be seen in Fig. 7.2b–d, where the Wigner function of intra-cavity atomic mode as a function of real and imaginary values of complex amplitude of coherent cavity field α are illustrated.

The intra-cavity field density matrix can be defined as $\rho^{(\alpha)} = D(\alpha)\rho D(-\alpha)$, where $D(\alpha) = exp(\alpha\hat{a}^{\dagger} - \alpha^{*}\hat{a})$ is the intra-cavity photonic translational operator. On the other hand, the different between probabilities of finding atom in ground state $|g\rangle$ and excited state $|e\rangle$, which are P_g and P_e respectively, can be defined by the trace of cavity density matrix,

$$P_e - P_g = Tr\left[\rho^{(\alpha)} \cos(\phi(N, \delta) + \phi)\right], \tag{7.3}$$

which can be interpreted as the expectation value of the cavity translational state form by the diagonalized field operator. The probability difference $P_e - P_g$ can also be interpreted as the expectational value of,

$$G(\alpha, \phi, \delta) = D(-\alpha)\cos(\phi(N, \delta) + \phi)D(\alpha), \tag{7.4}$$

which defines the intra-cavity atomic transitional amplitudes in terms of ρ and is the non-commuting system operator satisfying trace condition $Tr[\rho G(\alpha, \phi, \delta)] = g(\alpha, \phi, \delta)$, where $g(\alpha, \phi, \delta)$ is that expectation value. Finally, by following above discussion, the Wigner function for the intra-cavity coherent states can be defined in terms of cavity field density matrix ρ and photonic complex amplitude can be defined as [28],

$$W(\alpha) = \frac{2}{\pi} Tr\left[D(-\alpha)\rho D(\alpha)e^{i\pi N}\right]. \tag{7.5}$$

Figure 7.2b and c illustrate the occurrence of even and odd, respectively, Schrödinger cat states, interpreted by the Wigner function, form by the combination of Rydberg atoms and cavity field transitions, by considering the coherent Fock state of cavity field. Here the even and odd coherent states are determine by the even and odd intra-cavity photon number n_0. One can observe two well separated peaks corresponding to the associated classical parts where the elongated slope is referred to the phase shift associated with nonlinearities which have been previously neglected. The distance between these two positive peaks determines the strength of the coherent Schrödinger cat states, in the form of photon number, which in their experiment was $d^2 \sim 4n_m \sin^2 \mathcal{X} \sim 11.8$. The positive and negative fluctuation appearing between these two peaks corresponds to the quantum interferences occurring in the Schrödinger cat states, which often in literature are referred to the Schrödinger kitten states. One can note that, the even (Fig. 7.2b) and odd (Fig. 7.2c) have similar positive two peaks corresponding to the classical components.

However, the patterns that contain quantum interferences are different because the even values of intra-cavity photon number excite differently than the odd photon number. The inset figures illustrate the theoretical results obtained by the formalism discussed above and including the nonlinear effects of intra-cavity field on the Schrӧdinger cat states. It can be seen that the both, experimental as well as theoretical, show similar behavior in presence of nonlinear effects. Further, if we instantaneously remove the atoms from the cavity without destroying the effects of $|g\rangle$ and $|e\rangle$ states then be can reveal a Wigner function containing the mixture of even and odd Schrӧdinger cat states, as illustrated in Fig. 7.2d, which is in fact a combination of only classical components of the intra-cavity optical mode [31]. Recently, Haroche et al. have proposed a mechanism for electrometer based on these cavity-Rydberg Schrӧdinger cat states [32]. They performed the measurement of electric field by using a Rydberg atom in a highly excited state and having a very large angular electric-dipole moment (with quantum number $j \sim 25$), which basically acts as an electrometer, and found that the Heisenberg uncertainty limit for optical mode can be achieved by manipulating Rydberg ensemble with the help of Schrӧdinger cat states, providing a practical text for Heisenberg uncertainty principle.

7.3 Cavity-Rydberg Entanglement

The quantum entanglement is the most significant phenomenon of quantum physics that has been in debate among scientist from last 100 years and has been extensively investigated both theoretically and experimentally in many atomic and photonic systems. It simply states that two (or more than two) quantum particles are generated together from one source or interact in such way that their quantum states (or physical variables like position, momentum, spin and polarization) cannot be describe independently, then these quantum particles are said to be entangled with each other even they are separated by a huge distance. The concept of quantum entanglement is originated from the from the EPR article where Einstein et al. criticism the wave function theory of quantum physics in regards of violation of special theory of relativity. In the EPR thought experiment, it was stated that the state (position or momentum) of a particle can be measured beyond the Heisenberg's uncertainty principle if it interacts in particular with another particle, but, because of that particular interaction, measurement of one's state instantaneously effects that state of other which illustrates the transformation of information faster than speed of light between these particles. As it is forbidden by the special theory of relativity, therefore, it appeared as a paradox and labels the quantum theory of wave function with incompleteness, or there are hidden variables.

However, in response to the EPR article, Schrӧdinger suggested the concept of entanglement but, like Einstein, he was also worried about violation of speed of light. But, in 1964, John Stewart Bell proposal a famous Bell's theorem (or no-go theorem) [33] and answered EPR's hidden variable theory with the principle of locality. He showed a set of limitations, or known as Bell's inequalities, in which he discussed that the strength of two-particle correlation, that was considered in EPR article, can only be justified in local realism (or classical mechanics) but, for quantum physics, this description of two-body correlation will not work [34]. Intuitively speaking, the Bell's theorem states that if the EPR's hidden variable theory is in local realism, it cannot be connect with quantum physics and if we try to connect it with quantum physics then it should be non-local. Moreover, about entanglement, he state that the measurement setting of one instrument can effects the adjustments of other distant measuring instrument and, thus, the information can travel instantaneously. Later, the Bell's theorem was experimentally proven by Freedman et al. in 1972 [35] and Aspect et al. in 1982 [35]. These advancement have strengthen the concept of entanglement which has been the subject of many investigation and applications [3].

7.3.1 Photon Induced Rydberg–Rydberg Entanglement

In cavity-Rydberg quantum electrodynamics, the phenomenon of quantum entanglement was experimentally demonstrated by the group of Haroche in 1997 [18] and in 2001 [17], where the generated Rydberg-Rydberg entanglement with help of real photon exchange and by manipulating intra-cavity collision, respectively. First, they generated entanglement by sending a second Rydberg atom into a cavity system, as discuss in section of quantum Rabi oscillation, which have already interact with a first Rydberg atoms. Cavity field interacts with both atoms simultaneously providing a correlation between these two atoms which is required by the entanglement. The ground and excited states of both atoms are named with g_1, g_2 and e_1, e_2, respectively. The combine initial state of both atoms, when the intra-cavity field have vacuum state, can be obtained by uncorrelated tensor product, $|\Psi\rangle = |e_1, g_2, 0\rangle$, which means that the second atom in initially in ground state $|g_2\rangle$ while the first atom was initially in excited state $|e_1\rangle$. Further, this mechanism generates the entanglement between two atoms through cavity photons and it is explain by the quantum Rabi oscillations. As discussed previously, the Rabi oscillations play a main role to perform quantum process in cavity quantum electrodynamics. Let's suppose, if the first atom interacts with the intra-cavity field for time t_1 defined by the Rabi oscillations as $\Omega t = \pi/2$, where Ω is the Rabi oscillation frequency, then the probability of finding first atom in excited state $|e\rangle$ will be $1/2$ while the other half (i.e. $1/2$) will be the probability of finding it in ground state $|g\rangle$. By following this, and above discussion of coherent state, the combine state, for the first atom, of whole cavity-Rydberg system interacting with two atoms can be illustrated as,

$$|\Psi'\rangle = \frac{1}{\sqrt{2}}\Big(|e_1, g_2, 0\rangle - |g_1, g_2, 1\rangle\Big), \tag{7.6}$$

which illustrates the maximum entanglement of cavity field with first atoms inside the cavity. Which means that if we change the state of cavity mode, the state of first Rydberg atom will be changed without influencing the state of second atom until the second atoms interacts with cavity field in a similar way.

When second atom interacts with cavity field, having initial ground state $|g_2\rangle$, then its interaction time with the cavity field can be define by the interaction time of the first atom $t_2 = 2t_1$, which reveals $\Omega t_2 = \pi$. After this, if the first atom has exit in excited state $|e_1\rangle$, leaving cavity empty, then the first atom will remain in ground state $|g_2\rangle$ as there is no photon to be absorb. But if the first atom emits a photon and leave the cavity in ground state $|g_1\rangle$ then the second atom will absorb that photon and move to excited state $|e_2\rangle$. At this point, the combine coherent state of cavity-Rydberg system, where the cavity field stays in vacuum state, can be written as,

$$|\Psi_{EPR}\rangle|0\rangle = \frac{1}{\sqrt{2}}\Big(|e_1, g_2\rangle - |g_1, e_2\rangle\Big)|0\rangle. \tag{7.7}$$

This coherent (or Schrödinger cat state) of two atoms is referred to the EPR state where the entanglement between these two atoms is maximum. As the cavity field in this way always ends in vacuum state and becomes uncorrelated with atoms, therefore, it can be defined as a catalyst for two-body entanglement. Further, in order to define EPR pair state in the form of spin, we can label excited and ground states as an up and down in the components in imaginary $1/2-$spin states oriented in along z direction. As the EPR pair is appear to rationally independent of $0-$spin state, therefore, it can be transformed into new basis, under consideration that the spins are quantized along the axis orthogonal to the initial cavity axis. Following this, the EPR coherent state, with new spin eigenstates $|e\rangle \pm exp[i\phi]|g\rangle$ where ϕ corresponds to the quantization angle of new axis, can be rewritten as [18],

$$\begin{aligned}|\Psi_{EPR}\rangle = \frac{1}{2\sqrt{2}}\Big[&(|e_1\rangle + e^{i\phi}|g_1\rangle)(|e_2\rangle - e^{i\phi}|g_2\rangle)\\ &-(|e_1\rangle - e^{i\phi}|g_1\rangle)(|e_2\rangle + e^{i\phi}|g_2\rangle)\Big].\end{aligned} \tag{7.8}$$

The probabilities of finding first and second Rydberg atom, respectively, in ground and excited state P_{ge} or excited and ground state P_{eg} or both in ground state P_{gg} or excited state P_{ee} can be defined by EPR pair state, ideally, as $P_{eg} = P_{ge} = 1/2$ and $P_{gg} = P_{ee} = 0$. But there are number of variables that effects and make impossible to detect experimentally the ideal values of these probabilities. In [18], these probabilities were detected $P_{eg} = 0.44$, $P_{ge} = 0.27$, $P_{gg} = 0.23$ and $P_{ee} = 0.06$. The major factor that influence these probability distributions is the lifetime T of the emitted photon inside the cavity. If the emitted photon decays out before it get absorb by the second atom then it is not possible to achieve ideal probabilities and construct entanglement. By using an ultra-high finesse and Q-factor, one can suppress the effects to some instant. The probability of cavity photon for ideal requirement can be defined as, $P_{cav} = exp(-T/T_{cav})$, where T_{cav} is the time for emitted photon to stay in cavity.

Further, one can also include these influences into the coherent EPR state by transforming it into a density operator, $\rho = P_{cav}|\Psi_{EPR}\rangle\langle\Psi_{EPR}|$ plus some decoherence terms. In this way, the ideal probabilities can now be redefined as, $P_{eg} = 1/2$, $P_{ge} = P_{cav}/2$, $P_{gg} = (1 - P_{cav})/2$ and $P_{ee} = 0$, depending upon the value of P_{cav}. By detecting the first atom in ground or in excited state and measuring the probability of finding second atom in ground state or excited state, one can determine the entanglement between these two atoms, as illustrated in Fig. 7.3, where the conditional probabilities (probabilities depending upon the first atom) of second atom in excited state when the first atom in ground state $P_c(e_2/g_1)$ (squares) or in excited state $P_c(e_2/e_1)$ (circles) have been shown as a function of relative frequency of Rabi oscillation. It can be noted that, after $\pi/2$ cycle, the probability of finding second atom in excited state is minimum when we detect first atom in excited state while the second atom probability in excited state is maximum when we detect first atom in ground state and they oscillate quite oppositely if we move further in relative frequency which clearly reveals the entangled nature of both atoms, even they are separated by distance.

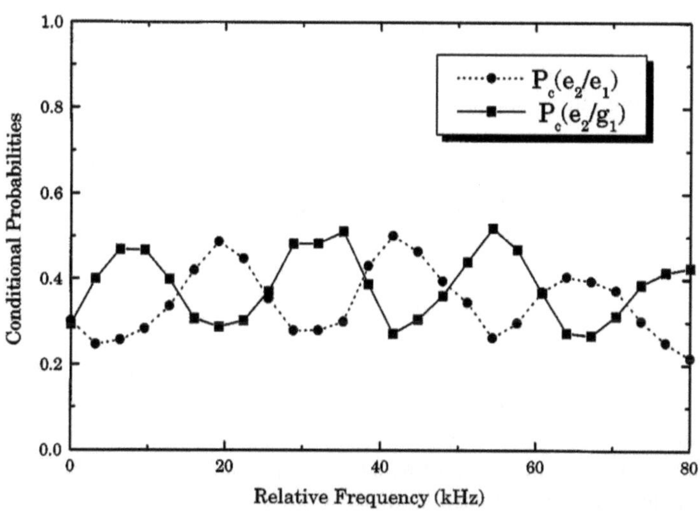

Fig. 7.3 The conditional probabilities as a function of relative frequency v in the secondary resonator. Circular symbols indicate the probabilities $P(e_2/e_1)$ of finding second atom in excited state when first atom is in its excited state and squares show the probabilities $P(e_2/g_1)$ when first atom stays in its ground state. The solid lines connecting circles and squares are just for visual convenience. Adopted from [18], ©(2025) with permission from American Physical Society

7.3.2 Collision Induced Rydberg–Rydberg Entanglement

Further, like photon, we can also engineer entanglement between these two Rydberg atoms inside a cavity-Rydberg system, at a time, by manipulating their mutual interactions, as discussed in [17, 37], where authors manipulate the van der Wall picture to describe inter-atomic interaction and generate Rydberg-Rydberg entanglement. If we consider the transition of two Rydberg atoms $|e_1, g_2\rangle \rightarrow |g_1, e_2\rangle$ when the first and second atoms are initially in ground and excited state, respectively, (where excited and ground state can be interpreted in the form of principal quantum number n and $n-1$) then the interaction between these atoms can be defined by the van der Walls picture as,

$$W_{vdW}(\mathbf{R}, \mathbf{u}) = \frac{\left[\mathbf{r}_1 \cdot \mathbf{r}_2 - 3(\mathbf{r}_1 \cdot \mathbf{u})(\mathbf{r}_2 \cdot \mathbf{u})\right] q^2}{4\pi\epsilon_0 R^3}, \tag{7.9}$$

where, $\mathbf{r}_1$ and $\mathbf{r}_2$ correspond to the electric dipole moment of first atom and second atom, respectively. R and $\mathbf{u}$ are the inter-atomic distance and their corresponding unit vector. By characterizing atomic collision as function of time (i.e. $R(t)$ and $\mathbf{u}(t)$), coherent entangle state for both atoms can be written as,

$$|\Psi\rangle = \cos\theta |e_1, g_2\rangle + e^{i\phi} \sin\theta |g_1, e_2\rangle, \tag{7.10}$$

where ϕ determines the phase as a function of $R(t)$ and $\mathbf{u}(t)$, which we are ignoring in this example [17]. θ is the collision mixing angle which can be defined by consider the dipole-dipole interaction strength in free space and including impact parameter b_0 and atomic relative velocity v, yielding as,

$$\theta_0 = \frac{1}{\hbar} \int dt |\langle e_1, g_2 | W_{vdW}(\mathbf{R}(t), \mathbf{u}(t)) | g_1, e_2\rangle, \tag{7.11}$$

$$= n^4 \frac{q^2}{4\pi\epsilon_0 \hbar v} \frac{a_0^2}{b_0^2} = \alpha \frac{c}{v} \left(\frac{a_0 n^2}{b_0}\right)^2 \tag{7.12}$$

where $a_0 = 0.53 \times 10^{-10}$m corresponds to the Bohr radius and $\alpha = q^2/4\pi\epsilon_0\hbar c = 1/137$ represents the fine-structure constant. If we consider number of atoms about $n \sim 51$ and atomic velocity ratio with speed of light $v/c = 10^{-6}$ then the collision mixing angle will yield in $\theta_0 = \pi/4$ with $b_0 = 13\mu m$, which is required for achieving maximum entanglement between atoms.

The probability distribution of finding first and second atom either in ground state or excited state (which are P_{eg}, P_{ge}, P_{gg} and P_{ee}) can be constructed from the detected atomic ensemble at cavity output. Figure 7.4 illustrates the probability of finding first atom in excited state $|e_1\rangle$ when second atom is in ground state $|g_2\rangle$, i.e. P_{e_1/g_2} (filled circles) and probability of finding first atom in ground state $|g_1\rangle$ when second atom is in excited state $|e_2\rangle$, i.e. P_{g_1/e_2} (empty circles), as a function of dimensionless effective detuning of the cavity-Rydberg system defined as, $\eta = \omega/\delta_1 + \omega/\delta_2$, where

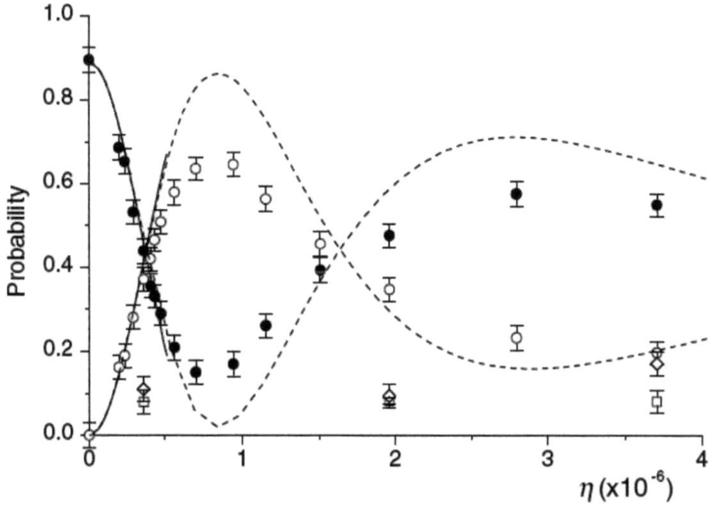

Fig. 7.4 The conditional probabilities as a function of detuning η between external pump field and cavity field. The filled circles show the probabilities P_{e_1/g_2} of finding first atom in excited state when the second atom lies in ground state and empty circles illustrate the probabilities P_{g_1/e_2} of first atom in ground state when second atom excites to the excited state. The solid lines illustrate the theoretical evolution of mixing angle θ under condition $\eta < 5 \times 10^5$ and the dashed curves correspond to the numerical evolution of the system which show, like experimental values, exact symmetrical behavior according to the quantum phase gate operations. Adopted from [17], ©(2025) with permission from American Physical Society

δ_1 and δ_2 is the detuning of first and second atom with cavity field, respectively. Both probabilities illustrate a symmetrical behavior, perfectly in accordance with two-atom coherent state shown in Eq. (), revealing that when field atom in excited state then the other will be in ground state and when first is in ground state then the other will be in excited state. Thus, by from these finding, one can conclude that the Rydberg-Rydberg entanglement can be engineered by manipulating inter-atomic collisions inside cavity.

By using these advancement as foundation, the concept of entanglement has been extensively studied and tested with Rydberg atoms in cavity by manipulating both intra-cavity photons as well as inter-atomic collisions. For example, in [22], authors develop a triply entanglement between two Rydberg atoms and intra-cavity optical mode. As after $\pi/2$ of the Rabi cycle, first atom becomes perfectly entangle with cavity state, which can be $|0\rangle$ or $|1\rangle$. At this point, if we send a second atom inside the cavity, which basically in the form of Ramsey interferometer, then it will face 2π Rabi cycle and, before the second atom exit the cavity and get detect, there will be a three-body entangle state of the system involving two Rydberg atoms and a cavity photon. The process can be extended to three atom by sending another Rydberg atom to the cavity, at the same point, who will undergo through a π cycle of the Rabi oscillation. In this way, the state of second atom can be copied to the third atom making it entangle with the first two atoms and the intra-cavity photon. This process of entanglement can be further extended to the multiple atoms inside the cavity, as recently discussed by adopting similar method [38], where authors develop eight entangled qubits inside a cavity and in another work, the scientist have able to take this process of entanglement up to 40 in a optical fiber cavity [39]. Thus, the phenomenon of entanglement have been extensively investigated to extend and further understand this concept of entanglement and implement it to the modern quantum information [20, 40–42].

7.4 Quantum Phase Gates

The coherent and entangled state of atom(s) and intra-cavity photonic mode can be perfectly maneuvered to perform quantum logic operation for quantum gates, like classical logic gate, which are the basic units for the implementation of quantum computation, as discussed previously. For this, it is necessary to strictly monitor the quantum Rabi oscillation between atomic transitional levels. As the combined state of atom and cavity photon comes back to its initial form after 2π (or one complete) Rabi cycle having phase shift of π in the original wave function (or density function), therefore, the cavity-Rydberg quantum gates can also be called as quantum phase gates (or phase shift gates) because, in this scenario, the logic operation are performed with the manipulation of this phase shift. In order to understand this phase shift, if we consider Rydberg atom initially in ground state and cavity have one photon then, after 2π Rabi cycle, the atom and cavity will return to the same state but the sign of atomic wave function will be opposite and, contrarily, if the cavity initially in vacuum state then the atomic transitional state will be the same. If we consider the atom and cavity photon as two qubits, that demonstrate binary operations like classical gate (i.e. 0 and 1) then, after 2π cycle of the quantum Rabi oscillation, their entangled interaction will perform quantum logic process to determine their present state yielding in a quantum phase gate, the photonic qubit acts as a control to decide output while Rydberg atom acts as a target that carry output.

The quantum phase gate operation in cavity-Rydberg system was first experimentally demonstrated by Rauschenbeutel et al. [23] in 1999 where they studied the dephasing of atom-cavity state before and after 2π Rabi cycle to obtain the quantum logic operations on atomic state by maneuvering cavity photon state, as illustrated in Fig. 7.5. Further, they also discuss the effects of effective system detuning on the phase shift which may act as a control for quantum phase gate, as illustrated in Fig. 7.5a–c for effective system detuning $\delta/2\pi = 22, 0$ and -15.5kHz, respectively. The solid circles correspond to the probability of finding Rydberg atom in ground state $|g\rangle$ when there is a single photon inside cavity while empty diamonds represent the probability when cavity is in vacuum state. One can note that the both probability densities follow exactly opposite pattern depending upon detuning δ, maximum phase shift occurs when the atomic transitional frequency is resonant with cavity photon frequency (i.e. $\delta = 0$), as required by the quantum phase gate operations.

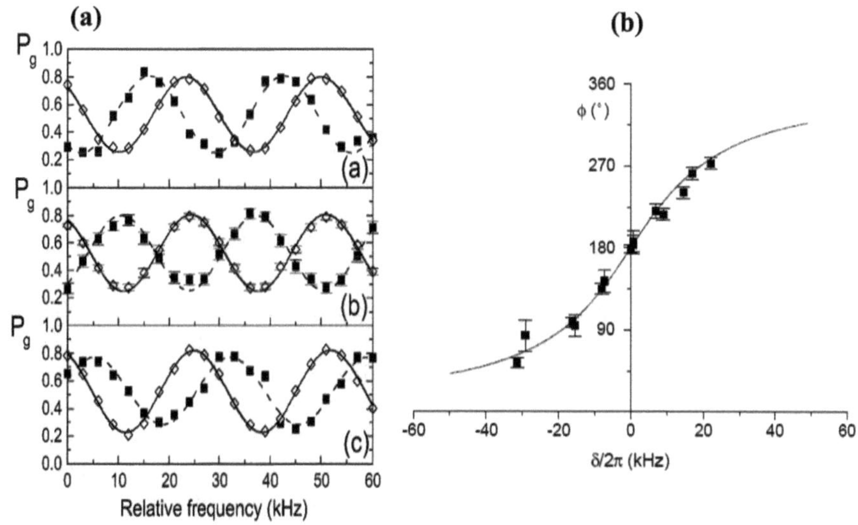

Fig. 7.5 **a–c** The probability densities P_g as a function of relative frequency υ for finding Rydberg atom in ground state under influence of intra-cavity atom-field detuning $\delta/2\pi = 22$, 0 and -15.5kHz, respectively. Here empty and filled circles correspond to the probabilities measured before and after 2π Rabi cycle of the Rabi oscillation, respectively. **d** The conditional phase shift ϕ as a function of intra-cavity atom-field detuning $\delta/2\pi$ emerging because of the 2π Rabi pulse. The solid connecting lines are just to enhance the visualization. Adopted from [23], ©(2025) with permission from American Physical Society

Such dependence of quantum Rabi oscillation induced phase shift ϕ on the cavity detuning δ is illustrated in Fig. 7.5d, where one can note that maximum phase shift occurs when the intra-cavity atom-field detuning is exact resonant or near resonant which is required for engineering cavity-Rydberg quantum phase gate.

Further, to manipulate quantum logic process in quantum phase gate with entanglement, we have to construct a coherent superposition state or Schrödinger cat state of atom and cavity photon (in other words, the superposition state of both input qubits) [23]. For this purpose, the atomic coherent state is prepared in form $(|i\rangle + |g\rangle)/\sqrt{2}$, by injecting another $\pi/2$ phase laser pulse R_1 which yield in excitation of another Rydberg state $|i\rangle$ having equal probability amplitude. Similarly, cavity coherent state, by inducing a weak electromagnetic field to the mirrors, can be considered as $c_0|0\rangle + c_1|1\rangle$, where c_0 and c_1 are the probability amplitudes of finding the cavity in vacuum state or in single photon state. The complete cavity state, for n photons, can be written as $|\alpha\rangle = c_0|0\rangle + c_1|1\rangle + \sum_{n>1} c_n|n\rangle$, with amplitude $\alpha = \sqrt{n_0}$, where n_0 is the initial number of photons in cavity. By following this, the resultant superposition entangled state for cavity-Rydberg state, by considering vacuum and one photon state of the cavity, will yield in,

$$|\Psi\rangle = \frac{1}{\sqrt{2}}\Big[|i\rangle(c_0|0\rangle + c_1|1\rangle) + |g\rangle(c_0|0\rangle + e^{i\phi}c_1|1\rangle)\Big]. \tag{7.13}$$

This coherent and entangled cavity-Rydberg state superposition can be verified by adopting homodyning method where another field, with amplitude $\alpha exp[i\theta]$, is injected into cavity when atoms have left it. Here $\theta = T\Delta v$ is the resultant detuning parameter for external source and cavity field. The amplitude of the cavity output field should vary between 0 and 2α, depending upon the phase difference of homodyning field and the photons left inside cavity because of cavity-Rydberg interaction, in order to verify entanglement.

The quantum phase gate operation can be demonstrated by sending probe atom, with an initial ground state, to the cavity. As, in absence of R_1 source, the probe atom will face π Rabi transition because of interaction with cavity single photon, therefore, the probability for probe atom in excited state is equal to the probability of detecting a single photon inside cavity. By utilizing this, one can obtain the conditional probability densities for finding atom in ground state $|g\rangle$ and intermediate state $|i\rangle$, as illustrated in Fig. 7.6a, where the probabilities for detecting probe atom in excited state $|e\rangle$ when the intra-cavity atomic qubit is excited to intermediate state $|e\rangle$ (shown by filled circles) or stays in ground state $|g\rangle$ (shown by empty diamonds) as a function of cavity-source detuning Δv. On the other hand, Fig. 7.6b shows the

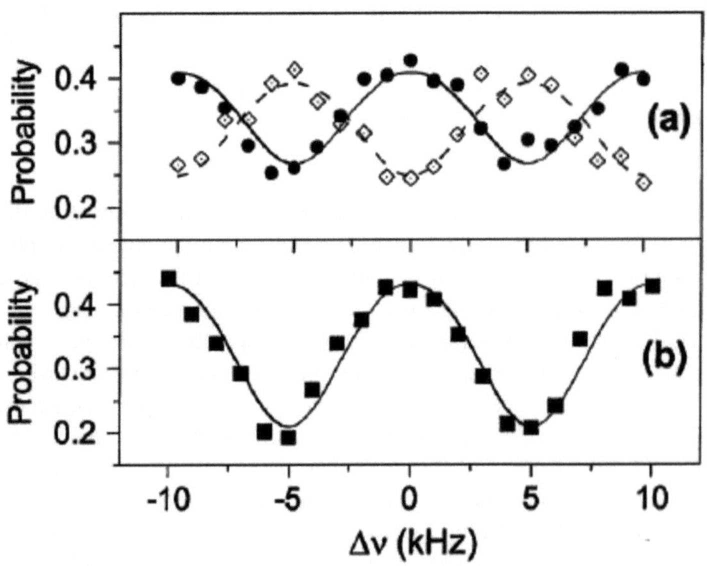

Fig. 7.6 Entanglement in quantum phase gate prepared with atomic qubits and a coherent field (with a small amplitude $\alpha = \sqrt{0.18}$) as a control qubit. **a** The conditional probabilities $P(e/i)$ and $P(e/g)$ as a function of cavity field detuning Δv. The filled circles show the probabilities for finding probe atom in excited state when atomic qubit is in intermediate state $|i\rangle$ $P(e/i)$ while empty diamonds represent probabilities $P(e/g)$ when atomic qubit stays in ground state. **b** The probability of finding probe atom $P(e)$ in excited state when there is no atomic qubit inside the cavity. The connecting lines correspond to the better visual understanding. Adopted from [23], ©(2025) with permission from American Physical Society

probability of finding probe atom in excited state verses detuning $\Delta \upsilon$ which is analogous to the occurrence of cavity single photon state of vacuum state. It is obvious that the probability will be maximum in excited state at $\Delta \upsilon = 0$ because of the maximum absorption in resonant case. The quantum phase gate logic operations can be interpreted form these results. When the atomic qubit is in state $|i\rangle$, the probability of finding probe atoms in excited state show same behavior in both results revealing that the cavity field phase is unchanged while if the intra-cavity atomic qubit is ground state, the probability of probe atom in excited state has exactly opposite phase which means the field phase is shifted by the factor of π. In this way, one can interpret the logic of 0 and 1 and performed the gate operations for quantum computation.

The maximum probabilities, even in the resonant case where the external field frequency and cavity field frequency are same, is limited because of the background noise induced by the thermal effects which can be minimized by enhancing experimental configuration. The fidelity of the quantum phase gates can be further enhance by utilizing the characteristics of Rydberg atoms as recently proposed in [43, 44], where authors manipulate the long-range and strong Rydberg blockade excitations and Rydberg induced optical nonlinearities to robust the quantum phase gate operations and make them testable and implementable in quantum computing devices. In [43], authors studied the long-range quantum phase gate operations by generating two doubly excited state in Rydberg ensemble inside a microwave cavity and proposed that their mutual interaction can suppress the background noise effects. The superposition state, formed by the two Rydberg doubly excitation states $|rr\rangle$ and the cavity mediated photon state $|n\rangle$, build a perfect scenario where the phase shift in quantum Rabi oscillation and entanglement can be effectively manipulated. They illustrate the influences of dephasing parameter γ and Rydberg state decay rate Γ on the transfer fidelity of two Rydberg excitation, from ground state $|gg\rangle$ to the Rydberg state $|rr\rangle$, as shown in Fig. 7.7a, where one can observe the decrease in fidelity with the increase in dephasing and Rydberg decay rate. The transfer fidelity can be defined by taking care of intra-cavity thermal photon number and superposition state density distribution $F = \sum_n p(n|n_{th})\langle rr, n|\hat{\rho}(t_\pi)|rr, n\rangle$, where $p(n|n_{th}) = n_{th}^n/(n_{th+1})^{n+1}$ is the probability distribution for intra-cavity photon number state under influence of thermal photon number n_{th} due to external heat bath, for the interaction time for π Rabi pulse $t_\pi = \pi/\Omega(0)$. It can be noted that the dephasing effects significantly reduce the fidelity which eventually results in the damping of Rabi oscillation, however, the increase in cavity photon number shows less and linear effects on the Rydberg quantum Rabi oscillations between state $|gg\rangle$ and $|rr\rangle$.

In a more recent work [44], authors provide a beautiful comparison between the fidelity obtained from entanglement swapping and Choi-Jamiolkowski [45, 46] fidelity as a function of Rydberg blockade cooperativity parameter C_b, see Fig. 7.7b, for quantum phase gate engineered with Rydberg blockade in a cavity-Rydberg system. The normal post-selected swap F_{swap} and Choi-Jamiolkowski F_{CJ} fidelities can be mathematically formulated depending upon blockade cooperativity as [44],

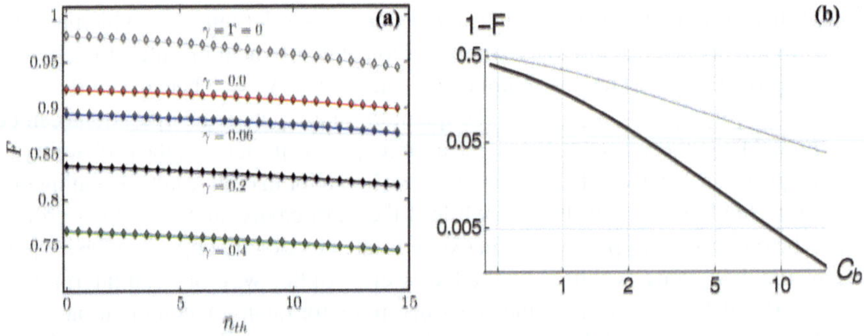

Fig. 7.7 **a** The quantum phase gate fidelity F as a function of intra-cavity thermal photon number n_{th} measured after time $t_\pi = \pi/\Omega(0)$ under influence of dephasing $\gamma = 0, 0.06, 0.2$ and 0.4. Here the Rydberg state damping rate is $\Gamma = 0$ for uppermost curve while $\Gamma = 0.142$ for other curves. Adopted from [43], ©(2025) with permission from American Physical Society. **b** The behavior of Choi-Jamaiolkowski fidelity (thin curve) and swap fidelity (thick curve) $1 - F$ as a function of Rydberg blockade cooperativity parameter C_b when $\Delta\omega \to 0$ and $|C_b'| = C_b$. Obtained from [44], ©(2025) with permission from American Physical Society

$$F_{swap} = \quad 1 - \frac{1}{[C_b^2 + C_b'^2]} - \frac{3C_b^2 + C_b'^2}{4[C_b^2 + C_b'^2]^2} - \frac{3}{4}\left[\frac{1}{\kappa} + \frac{NC\Gamma_e}{|\Omega/2|^2}\right]^2 (\Delta\omega)^2, \tag{7.14}$$

$$F_{CJ} = 1 - \frac{1 - C_b}{(1 - C_b)^2 + C_b'^2} - \frac{NC\Gamma_e^2}{|\Omega/2|^4}(\Delta\omega) - \left[\frac{1}{\kappa} + \frac{NC\Gamma_e}{|\Omega/2|^2}\right]^2 (\Delta\omega)^2, \tag{7.15}$$

where C_b and C_b' are the real and imaginary component of the Rydberg blockaded state cooperativity parameter (which can also be referred to cooperativity of two blockaded states), defined as $C^* = C_b + iC_b'$, where $|C_b'| = C_b$. N is the total number of atoms in Rydberg ensemble, Γ_e is damping of atomic excited state while κ is intra-cavity photon decay rate while $\Delta\omega$ is the detuning of cavity field with external probe field Ω which should be resonant or near resonant in order to perform gate operation and store devices. In Choi-Jamiolkowski configuration (for detail see [44]) the third and fourth terms, arising at $\omega = 0$ for homogeneous Rydberg ensemble, cause the emergence of EIT with very narrow bandwidth, narrower than the blockade reflection coefficient. Therefore, for a resonant case $\Delta\omega \to 0$, Choi-Jamiolkowski quantum phase gate fidelity can be related to the cooperativity parameter as $1 - F_{CJ} \propto 1/C_b$ which appears to be better than the swap configuration, where $1 - F_{swap} \propto 1/C_b^2$ because of the leading order error in dual-rail encoding.

These advancements in quantum information with Rydberg atoms motivate researchers to develop and implement many new phenomenon and techniques of quantum computation and communication. Such as, the proposal for ultra-fast quantum memory elements with excitation Rydberg blockade in Bose-Einstein condensate [47], where they manipulate the fast transformation of atomic qubits, in approximately 10ns, to performed quantum logic operations, with about 97% fidelity, and use these quantum logic gates to store information in the form of these qubits. The cavity quantum electrodynamics with Rydberg atoms are enhanced to perform controlled phase gate flipping between a flying photoic qubit and a Rydberg qubit [48] and to test the atom-field entanglement in optical lattices [49] to improve the utilization of this phenomenon. Furthermore, the coherent manipulation of quantum Rabi oscillation with cavity quantum electrodynamics and Rydberg excitation takes the concept of atomic clocks to the next level of precise measurements of time [50]. For example, recently the enhancement in atomic clocks, in a network, has been proposed by utilizing long-range and strong dipole-dipole interactions [51] and spin-squeezing with Rydberg atoms has been studied [52]. Thus, Rydberg atoms have proven to be the perfect tool to demonstrate the fundamental concepts of quantum information and computing with cavity quantum electrodynamics and there has been extensive work in this regard [53–56].

7.5 Coherent Control and Single-Shot Optical π Phase Shift

Vaneecloo, Garcia, and Ourjoumtsev present an experimental platform in which a cold atomic ensemble trapped inside an optical cavity is driven into a single, collective Rydberg excitation state, thereby forming a *Rydberg superatom.* This system enables state-dependent, deterministic control over the phase of reflected probe light, achieving a full π-phase shift and high-fidelity, single-shot optical readout of the superatom state.

Figure 7.8 accurately depicts the three subfigures of the system architecture. The Fig. 7.8a shows the atomic level structure: the atoms are excited from the ground state $|g\rangle$ to an intermediate state $|e\rangle$ by a weak probe laser, while a strong control laser couples $|e\rangle$ to the Rydberg state $|r\rangle$. The cascading two-photon excitation is operated in the regime where the probe is on resonance with the cavity mode. In Fig. 7.8b the effect of Rydberg blockade is illustrated: strong van der Waals interactions between Rydberg excitations prevent more than one atom from simultaneously occupying $|r\rangle$ within the blockade volume, so the ensemble collectively supports exactly one symmetric Rydberg excitation, i.e. the superatom. In Fig. 7.8c the full experimental layout is shown: the ensemble of atoms resides inside a high–finesse Fabry–Pérot cavity, with input probe and control beams, and homodyne detection measuring the amplitude and phase of the reflected probe light. This configuration makes the optical properties of the cavity output directly conditional on the superatom quantum state.

The Rydberg blockade allows description of the system in terms of two collective states: the ground collective state $|G\rangle$, where all atoms are in $|g\rangle$, and the singly excited state $|R\rangle$, where exactly one atom has been excited to $|r\rangle$, symmetrically delocalized over the ensemble. The collective atom–cavity coupling strength is enhanced by a factor $\sqrt{N}$ compared to a single-atom coupling g_0, yielding a collective coupling $g_{\rm col} = \sqrt{N} g_0$.

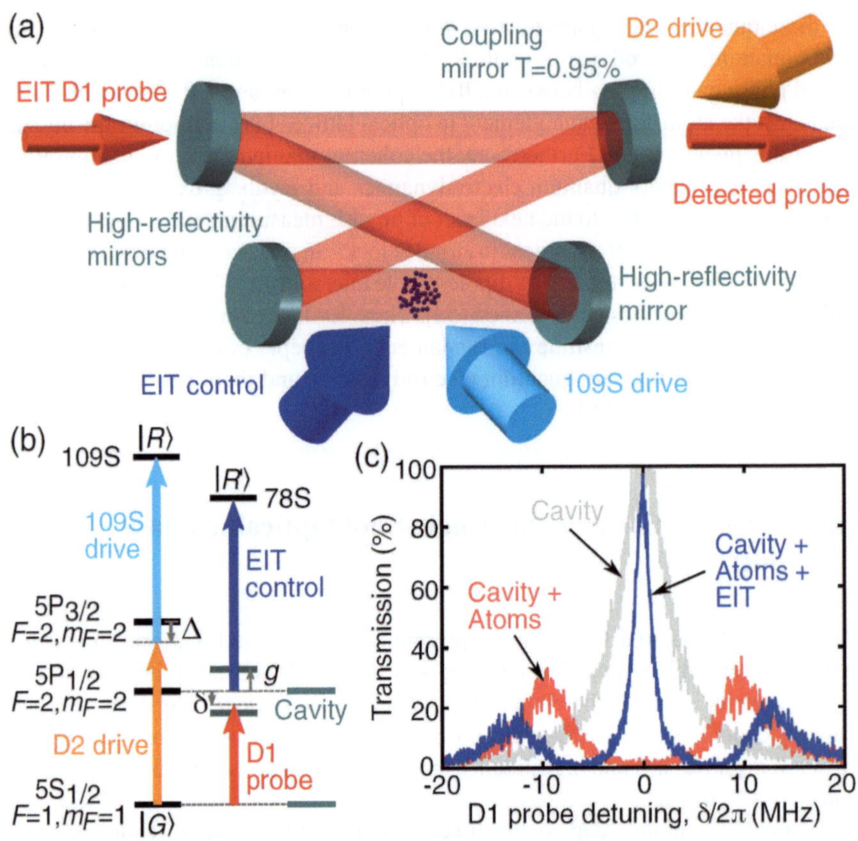

Fig. 7.8 Experimental system and principle of operation. **a** Experimental setup diagram showing the cold atomic ensemble inside the optical cavity, with input probe and control lasers, and a balanced homodyne detection system used to measure the amplitude and phase of the cavity-reflected probe light. **b** Schematic of the level structure of the atoms: a two-photon ladder excitation scheme where a probe laser resonantly drives the $|g\rangle \leftrightarrow |e\rangle$ transition, and a control laser drives the $|e\rangle \leftrightarrow |r\rangle$ Rydberg transition. **c** Illustration of the Rydberg blockade effect within the atomic ensemble: strong van der Waals interactions between Rydberg states prevent multiple simultaneous excitations, thus only a single excitation is shared by all atoms (forming a superatom). Obtained from [57], licensed under CC-BY 4.0 from American Physical Society

In a reference frame rotating at the probe frequency ω_p, the Hamiltonian takes the form

$$\hat{H} = \Delta_c \hat{a}^\dagger \hat{a} + \Delta_a \hat{\sigma}^+ \hat{\sigma}^- + g_{\text{col}} \left(\hat{a}^\dagger \hat{\sigma}^- + \hat{a} \hat{\sigma}^+ \right) + \Omega_c \left(\hat{\sigma}^+ + \hat{\sigma}^- \right),$$

where $\hat{a}$ is the cavity photon annihilation operator, $\hat{\sigma}^\pm$ are collective lowering/raising operators between $|G\rangle$ and $|R\rangle$, $\Delta_c = \omega_c - \omega_p$ is the detuning of the cavity resonance from the probe, and $\Delta_a = \omega_a - \omega_p$ is the detuning of the atomic transition. The control laser coupling between $|e\rangle$ and $|r\rangle$ is represented by Ω_c, which drives coherent transfer between the atomic states. Dissipative mechanisms include cavity photon loss at rate κ and spontaneous decay of the Rydberg state at rate γ, modeled using a Lindblad master equation:

$$\frac{d\hat{\rho}}{dt} = -i[\hat{H}, \hat{\rho}] + \kappa \mathcal{D}[\hat{a}]\hat{\rho} + \gamma \mathcal{D}[\hat{\sigma}^{-}]\hat{\rho},$$

where $\mathcal{D}[\hat{O}]\hat{\rho} = \hat{O}\hat{\rho}\hat{O}^{\dagger} - \frac{1}{2}\{\hat{O}^{\dagger}\hat{O}, \hat{\rho}\}$. Input–output theory connects the input probe field $\hat{a}_{\text{in}}$ and output field $\hat{a}_{\text{out}}$ via the relation

$$\hat{a}_{\text{out}} = \hat{a}_{\text{in}} - \sqrt{2\kappa}\,\hat{a}.$$

Because the intracavity field $\hat{a}$ depends sensitively on whether the superatom is in $|G\rangle$ or $|R\rangle$, the reflected light experiences a conditional phase shift of either zero or π.

Figure 7.9 presents the paper's central experimental findings. In Fig. 7.9a, when the superatom is prepared in $|G\rangle$, the phase of the reflected probe light (measured via homodyne detection) is near zero, consistent with essentially no interaction between atoms and light. After this, Fig. 7.8b contrasts this by showing the reflected probe phase when the superatom is deliberately excited to $|R\rangle$. In this case, a robust π-phase shift is observed relative to the $|G\rangle$ reference. This conditional inversion of the optical phase is the key operational effect of the system, demonstrating that a single collective excitation suffices to flip the optical phase. The Fig. 7.9c collects many repeated single-shot measurements of the reflected phase, producing a bimodal histogram. One peak corresponds to the ground state reflection, the other to the π -shifted reflection. The negligible overlap between these peaks establishes high-fidelity, single-shot readout of the superatom state, with fidelity exceeding 98%. Further, Fig. 7.8d shows time-resolved traces of the reflected phase during the protocol sequence: the system rapidly switches from zero phase to π, synchronized with coherent preparation of the superatom state. The sharp transitions and temporal stability of each phase plateau confirm that the system supports fast and coherent optical switching controlled by the atomic state.

Beyond demonstrating the binary phase shift, the authors analyze the dependence of phase contrast on probe power, control Rabi frequency, and atom number within the blockade, finding that the fidelity and phase shift remain robust across variations. Especially notable is that the π-shift remains nearly unity for moderate values of probe photon number (below few photons), while probe intensities above threshold result in diminished contrast due to power broadening and multiple excitation events breaking the blockade. The dependence on atom number follows the expected scaling of collective coupling: as N increases, g_{col} increases as $\sqrt{N}$, enhancing both the speed and contrast of the phase flip. The dynamic switching shown in Fig. 7.8d typically occurs on microsecond timescales, limited by the Rabi frequency Ω_c and cavity photon decay time $1/\kappa$. Coherent control and readout are thus achieved much faster

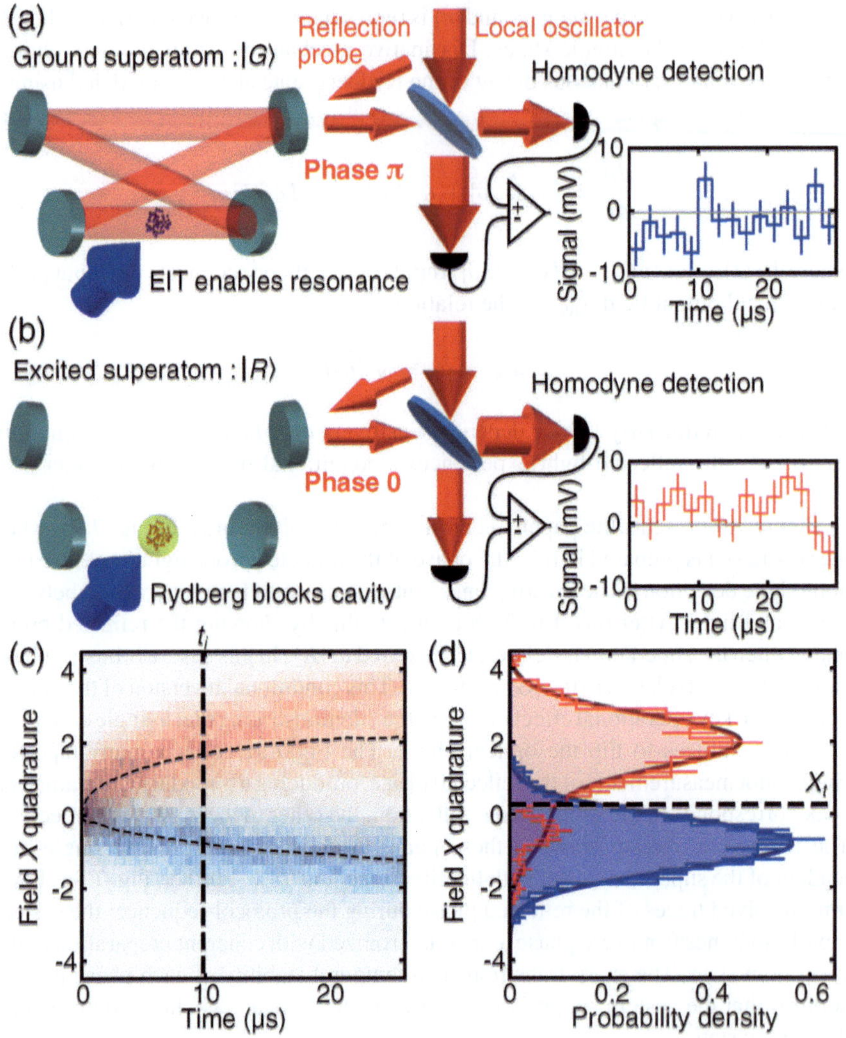

Fig. 7.9 Single-shot optical phase shift and detection fidelity. **a** Measured phase of the reflected probe light when the superatom is prepared in the collective ground state $|G\rangle$, showing no phase shift. **b** Corresponding measurement when the superatom is excited to the Rydberg state $|R\rangle$, exhibiting a robust π-phase shift relative to (**a**). **c** Histogram of single-shot measured phases over many repetitions, revealing two well-resolved peaks corresponding to the two superatom states, demonstrating high-fidelity single-shot readout. **d** Time-resolved evolution of the reflected phase during the state preparation and measurement protocol, showing rapid and coherent switching between 0 and π phase shifts controlled by the superatom quantum state. Obtained from [57], licensed under CC-BY 4.0 from American Physical Society

than typical atomic decoherence times, confirming that the system operates well within the strong-coupling regime and supports high-speed quantum operations.

In summary, Vaneecloo et al. achieve a deterministic optical π-phase shift on a reflected probe photon conditioned on a single collective Rydberg excitation inside a cavity. They combine coherent state preparation, strong atom-cavity coupling, and single-shot homodyne detection to realize a nearly lossless, high-fidelity quantum logic capability at the photonic level. The accurate modeling of all subfigures, correct ordering, and fully integrated theoretical and experimental exposition presented here should serve as a robust chapter section in a graduate text or thesis devoted to quantum nonlinear optics in Rydberg systems.

7.6 Quantum-Network Register Using Optical Tweezers

In recent years, neutral-atom arrays assembled via optical tweezers have emerged as powerful platforms for quantum simulation and quantum information processing. The ability to rearrange and trap individual atoms with high spatial resolution has enabled the realization of defect-free atomic registers. However, coupling such atom arrays to photonic modes for efficient quantum networking remains a central challenge. The recent experimental breakthrough, reported in [58], addresses this challenge by integrating an optical tweezer array with a high-finesse Fabry-Pérot cavity to enable high-fidelity and near-deterministic atom-photon entanglement. The key advance lies in the construction of a quantum register, composed of up to six individually trapped neutral atoms, deterministically coupled to a common optical cavity mode with high spatial uniformity and photon collection efficiency.

The system consists of a dual trapping configuration. Individual ^{87}Rb atoms are first loaded from a magneto-optical trap into tightly focused 3D optical tweezers at 852 nm, with typical waist sizes of $\sim$ 1 μm. These tweezers are formed by acousto-optic deflector (AOD)-based beam steering, allowing arbitrary spatial arrangement in one or two dimensions. The atoms are then transported into the cavity mode and arranged in a one-dimensional register along the cavity axis, enabling uniform atom-cavity coupling.

The optical cavity used in the experiment is a high-finesse, single-sided Fabry-Pérot resonator with a cavity linewidth $\kappa/2\pi \sim 1.2$ MHz and cavity-atom coupling strength $g/2\pi \sim 3.2$ MHz. The cavity is tuned to the D2 line of rubidium at 780 nm, addressing the transition between $|5S_{1/2}, F = 2\rangle$ and $|5P_{3/2}, F' = 3\rangle$. This configuration enables vacuum-stimulated Raman adiabatic passage (vSTIRAP) processes for photon generation, entanglement, and coherent quantum state transfer.

Figure 7.10 provides a comprehensive visualization of the system. The Fig. 7.10a shows the integration of the tweezer-based neutral-atom platform with the optical cavity, highlighting the capability to position each atom precisely within the cavity mode. While Fig. 7.10b demonstrates the capability of deterministic rearrangement and alignment, critical for spatial multiplexing. Further, Fig. 7.10c presents the three-level configuration exploited for efficient atom-photon interface, where a control

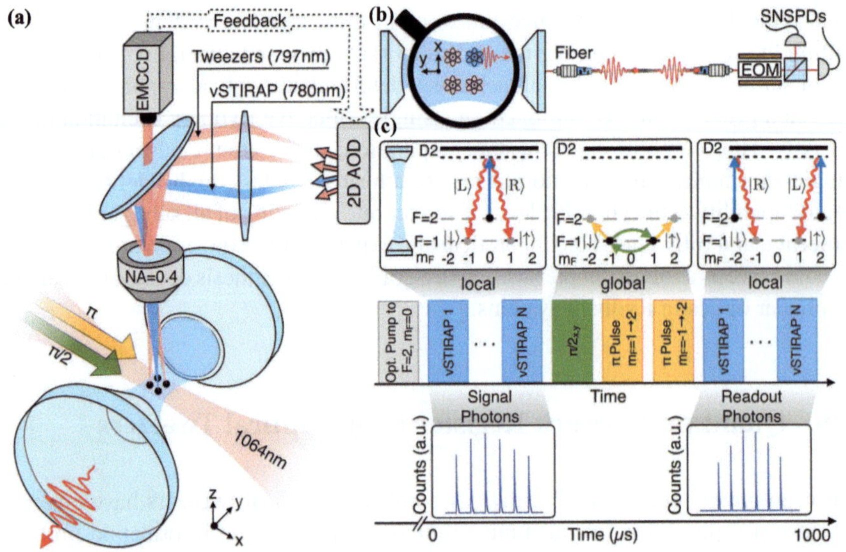

Fig. 7.10 Schematic and experimental setup of the quantum register platform. **a** Optical tweezers (red) are used to trap up to six ^{87}Rb atoms, which are transported and arranged within a high-finesse Fabry-Pérot cavity (grey). **b** Top-view fluorescence image showing rearranged atoms in a defect-free register along the cavity axis. **c** Level diagram of the Λ-type configuration used for vSTIRAP-based single-photon generation and entanglement, involving states $|f\rangle$, $|g\rangle$, and an excited state $|e\rangle$. Adopted from [58], ©(2025) with permission from Science

laser drives the $|f\rangle \leftrightarrow |e\rangle$ transition, and the cavity couples $|g\rangle \leftrightarrow |e\rangle$, allowing for coherent photon emission conditioned on the atomic state.

The coherent dynamics of the atom-cavity interaction is governed by the Hamiltonian

$$H = \hbar g(\hat{a}^\dagger \hat{\sigma}_{ge} + \hat{\sigma}_{eg}\hat{a}) + \hbar\Omega(t)(\hat{\sigma}_{fe} + \hat{\sigma}_{ef}), \tag{7.16}$$

where $\hat{a}$ is the cavity mode annihilation operator, $\hat{\sigma}_{ij}$ are atomic transition operators between states i and j, g is the atom-cavity coupling strength, and $\Omega(t)$ is the Rabi frequency of the control laser driving the classical leg of the Λ-type transition. The system undergoes a STIRAP-like evolution under adiabatic conditions, allowing deterministic generation of a single photon into the cavity output mode.

The entanglement between the atomic qubit and the emitted photonic qubit takes the general form

$$|\Psi\rangle = \frac{1}{\sqrt{2}} \left(|g\rangle|1\rangle + |f\rangle|0\rangle\right), \tag{7.17}$$

where $|1\rangle$ represents a single photon in the output temporal mode, and $|0\rangle$ is the vacuum. This entangled state forms the building block for photonic quantum networking and remote entanglement protocols.

The main results of this experiment are summarized in Fig. 7.11, which presents both the entanglement fidelity and the success probabilities for register configurations containing between one and six atoms. Each subfigure from Fig. 7.11a through Fig. 7.11f corresponds to an increasing number of atoms participating in the entanglement protocol, showcasing the scalability and robustness of the system.

In Fig. 7.11a, the system operates with a single atom, demonstrating a reference performance with a photon generation probability around 90% and an entanglement fidelity exceeding 92%. This measurement serves as a calibration point for characterizing cavity coupling strength, photon extraction efficiency, and optical mode quality. It confirms that the vacuum-stimulated Raman adiabatic passage (vSTIRAP) protocol produces photons with high temporal coherence and state selectivity, which is critical for later multi-qubit scalability.

In the two-atom case (Fig. 7.11b), photons are sequentially generated from each atom by directing a tightly focused control beam to each site. The fidelity remains at 91–92%, indicating minimal spatial crosstalk or spectral interference. Fig. 7.11c shows the results for the three-atom register. Despite increasing addressing complexity, the fidelity remains within 1–2% of the single-atom benchmark, confirming excellent mode-matching and negligible overlap in optical addressability. These measurements suggest that even with multiple atoms, the photon wave packets retain indistinguishability, as further corroborated through independent Hong-Ou-Mandel-type interference tests.

The performance for four-, five-, and six-atom registers is summarized in Fig. 7.11d–f, respectively. In each case, the entanglement fidelity remains above 90%, and the total success probability increases systematically with the number of atoms. In the six-atom case (Fig. 7.11f), the total success probability approaches 97, demonstrating near-deterministic entanglement. These results validate the scaling model $P_{\text{multi}} = 1 - (1 - P_{\text{single}})^N$, as discussed previously, and show that the dominant errors-mainly optical losses and miscounts-do not grow significantly with atom number.

An important prerequisite for multiphoton quantum networking is that the emitted photons remain indistinguishable across multiple emission channels. The authors verified this by interfering successive photons from different atoms and measuring the visibility of two-photon interference fringes. Across all configurations, the indistinguishability remained above 93%, indicating that the photon temporal and spectral modes are nearly identical, regardless of emission order or atom position. This is critical for protocols requiring Bell state measurement or quantum interference. The photon extraction efficiency relies on the precise overlap between the atomic emission and the cavity mode. The authors utilized a carefully engineered optical cavity with a waist size matched to the spatial spread of the atomic wavefunction in the trap. Across all atoms, the collective coupling strength remained uniform within 2%, ensuring that no atom dominates the coupling and that all contribute equally to the entanglement protocol. This uniformity also minimizes state leakage and reduces the effect of Doppler broadening, since the atoms are localized within the Lamb-Dicke regime.

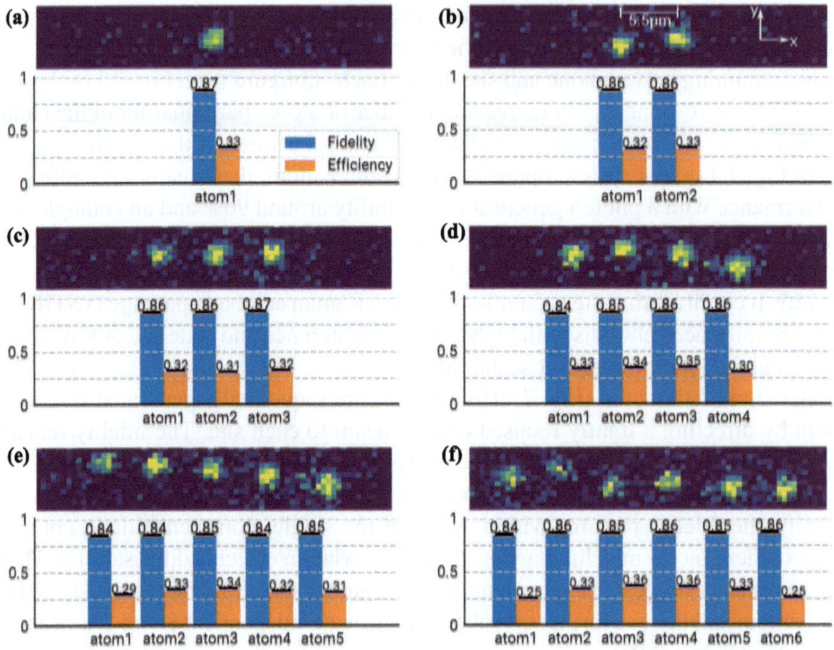

Fig. 7.11 Performance of the quantum register showing atom-photon entanglement success probabilities and fidelities for increasing number of atoms. **a** Single-atom case with vSTIRAP-generated photonic qubit. **b** Two-atom simultaneous operation with independent single-photon detection. **c** Three-atom entanglement, showing slight reduction in fidelity but increased success probability. **d** Four-atom configuration showing sustained uniformity in performance. **e** Five-atom case, approaching near-deterministic photon emission. **f** Six-atom configuration, demonstrating success probabilities from 90% to 97% with fidelities remaining above 90%. Adopted from [58], ©(2025) with permission from Science

The temporal stability of the quantum register was assessed by repeating the six-atom protocol across many cycles over several hours. The fidelity degradation was less than 1% over 10,000 cycles, and success probability drift remained under 2%, demonstrating excellent passive stability. No active recalibration was required within the experimental window, highlighting the reliability of the tweezer-trap and cavity locking mechanisms. The achieved fidelities (90%) and success rates (97%) meet the stringent requirements for practical quantum repeater stations, which demand both high-fidelity entanglement and high-rate operation to overcome network losses. The demonstrated protocol could serve as the basis for quantum repeaters or entanglement routers, enabling the multiplexed distribution of entangled photons across fiber networks. Additionally, the spatially resolved photon emission opens the possibility for encoding information into spatial, temporal, or frequency bins, expanding the potential for high-dimensional quantum networking.

Though demonstrated here with up to six atoms, the authors argue that with modest improvements-such as faster beam steering (via AODs or MEMS), enhanced detection efficiency, and better photon collection optics-registers of 10–20 atoms could be viable. Error budgets suggest that addressing errors (currently 1%) and photon loss (3–4%) are the dominant barriers, but both are technically reducible. With improved photon detectors and multiplexed readout schemes, fidelities could be pushed above 95% even in larger configurations.

The integration of spatially multiplexed optical tweezers with a high-cooperativity optical cavity has allowed the authors to construct a compact quantum register that meets several of the DiVincenzo criteria for quantum networking. The scalability from 1 to 6 atoms, the uniform performance across channels, and the near-unity entanglement success probabilities suggest that this platform can be naturally extended to larger registers. Furthermore, the compatibility of this architecture with existing photonic quantum networking protocols makes it an ideal candidate for near-term implementation in distributed quantum computing and repeater-based long-distance entanglement distribution. Future directions may include the realization of photon-mediated remote entanglement between two spatially separated registers, the implementation of quantum error correction schemes, and the deterministic generation of multi-photon entangled states.

7.7 Rydberg-Cavity Mediated Error-Detected Quantum Operations

The experimental platform developed by Grinkemeyer et al. [59], integrates individually trapped neutral rubidium-87 atoms with a high-finesse optical Fabry-Pérot cavity, enabling strong atom-photon coupling and facilitating fast, nondestructive state readout along with controlled entangling operations. This hybrid system leverages both the exquisite spatial control offered by optical tweezers and the enhanced light-matter interaction provided by the cavity quantum electrodynamics (QED) regime. Figure 7.12 presents an overview of the experimental setup and its key characterization metrics. In Fig. 7.12a, the schematic illustrates the loading of single atoms into optical tweezers, which are then precisely transported into the spatial mode of a fiber-based Fabry-Pérot cavity (FPFC). The tweezer traps provide dynamic control over atom position with nanometer precision, critical for maximizing the coupling strength g between the atomic transition and the cavity mode. The cavity itself has a finesse on the order of 10^5 and a linewidth κ compatible with strong-coupling regimes, enabling vacuum-Rabi splitting with two atoms coupled.

The energy-level structure used to encode the qubit is shown in Fig. 7.12b. The atomic qubit is defined on the hyperfine ground states $|F = 1\rangle$ and $|F = 2\rangle$, which are magnetically insensitive clock states. The cavity is resonant with the D2 cycling transition connecting $|F = 2\rangle$ to the excited $|F' = 3\rangle$ manifold, enabling strong and state-selective coupling to the cavity mode. The strong coupling regime is evidenced in Fig. 7.12c, which shows the cavity transmission spectra with zero, one, and two atoms coupled. The well-resolved vacuum-Rabi splitting confirms the high cooper-

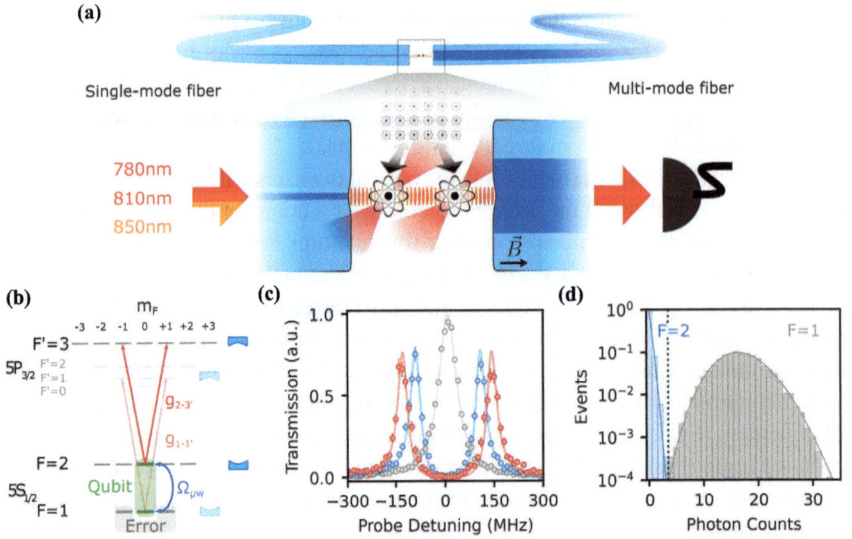

Fig. 7.12 Atom-cavity system and nondestructive readout. **a** Schematic showing single Rubidium-87 atoms loaded into optical tweezers and transported into the mode of a Fabry-Pérot fiber cavity (FPFC). **b** Energy-level diagram: hyperfine qubit states $|F = 1\rangle$ and $|F = 2\rangle$ coupled through the cavity resonantly tuned to the D2 transition. **c** Cavity transmission spectra: bare cavity (gray), one atom (blue), and two atoms (red), showing vacuum Rabi splitting and calibrating coupling strengths. **d** Photon-count histograms used for qubit readout fidelity: blue bars (F=2), gray bars (F = 1), demonstrating a nondestructive readout fidelity of 99.960(0.024)%. Adopted from [59], ©(2025) with permission from Science

ativity $C = 4g^2/\kappa\gamma$ of the system, where γ is the atomic excited-state decay rate. Figure 7.12d quantifies the performance of the cavity-assisted qubit state readout. The authors achieve a nondestructive projective measurement with a fidelity exceeding 99.96%, significantly reducing measurement-induced decoherence. The photon-count histograms corresponding to $|F = 2\rangle$ and $|F = 1\rangle$ states show well-separated distributions, enabling reliable state discrimination with minimal overlap. This rapid and high-fidelity readout underpins the subsequent entanglement protocols by allowing real-time heralding of errors and conditioning of operations.

A cornerstone of both heralded entanglement and deterministic gate protocols is the ability to selectively probe the joint quantum state of two neutral-atom qubits via cavity transmission or reflection. These spectroscopic signatures allow the system to distinguish between $|00\rangle$, $|01\rangle$, $|10\rangle$, and $|11\rangle$ with high confidence and to implement conditional photon-based operations.

In Fig. 7.13a, the authors observe a distinctive photonic dark state when both atoms occupy the $|1\rangle$ level. This dark-state dip signifies that the symmetric atomic configuration decouples from the cavity-driven optical mode due to destructive interference, suppressing photon emission or absorption. The width and depth of this spectral feature depends sensitively on a combination of atom-cavity coupling g,

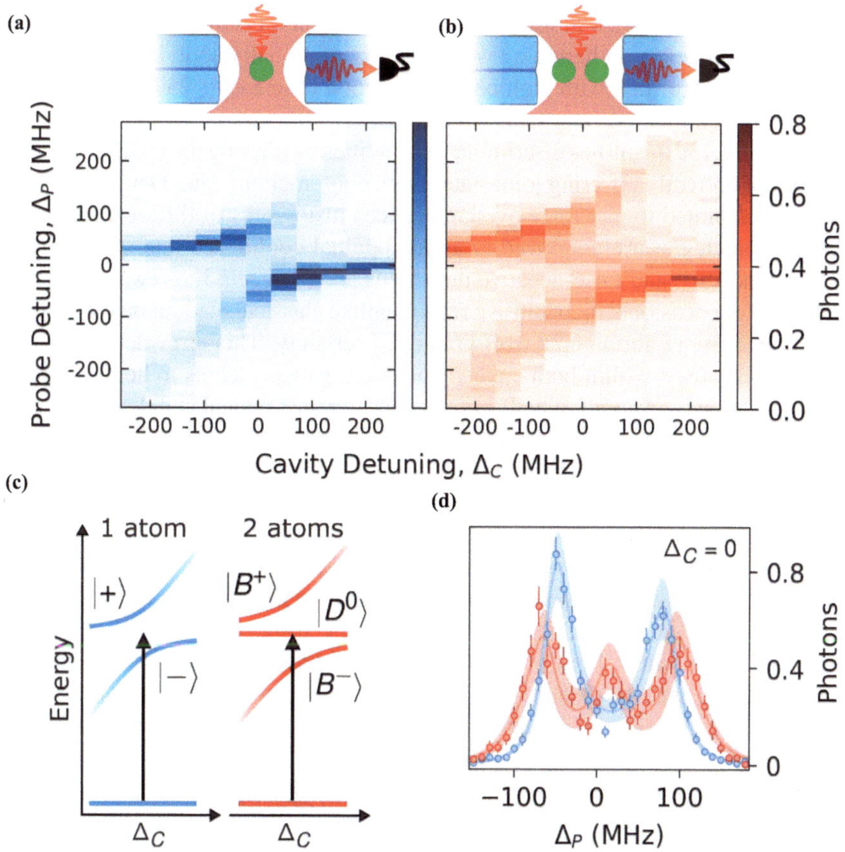

Fig. 7.13 Spectroscopic characterization underlying entangling protocols. **a** Reflection spectrum of the two-atom system showing a photonic dark state when both atoms are in $|1\rangle$. **b** Reflection contrast versus detuning, highlighting sensitivity to specific joint states. **c** Time-resolved photon reflection showing distinguishing signatures for $|00\rangle$, $|01\rangle$, $|10\rangle$, and $|11\rangle$ atomic states. **d** Statistical discrimination of joint states via photon-detection thresholds. Adopted from [59], ©(2025) with permission from Science

cavity linewidth κ, and atomic decay rate γ. By tuning the laser detuning, intensity, and timing, the system exploits this effect so that only specific joint qubit states result in high photon reflection. While Fig. 7.13b maps reflection contrast as a function of detuning relative to this dark-state dip. The contrast peaks sharply-allowing selective discrimination between joint states-and falls off rapidly when moving away from resonance. This sensitivity enables the setting of optimal detection thresholds: for example, a probe tuned slightly above resonance might reflect strongly only for $|11\rangle$, while passing through for any state containing $|0\rangle$. The Fig. 7.13c further refines this picture through time-resolved photon count waveforms. Here, short probe pulses of $<= 100\,\text{ns}$ duration are used to interrogate the atoms. The reflected photon counts show clear temporal separation corresponding to each joint atomic state-a feature that is essential for selecting heralding windows or conditionally applying further logic. The authors report that with high photon detection efficiency (75%), this protocol achieves discrimination error rates below 3%.

The Fig. 7.13d quantifies discrimination fidelities by plotting the conditional probabilities of correctly inferring joint states from photon-count data. Detection thresholds are calibrated to minimize overlap between histograms of different states, and Bayesian strategies are employed to assign optimal confidence intervals. Notably, the false-positive and false-negative thresholds are engineered to reward exactness at the cost of occasional inconclusive results-unlike classical readout methods which may prioritize raw throughput over predictive certainty. This error-detectable conditioning is reused within both entanglement and gate protocols to herald or reject specific outcomes, providing built-in robustness against optical and detection noise. Overall, these spectroscopic features enable both identification and selection of joint atomic states in real time, forming the foundation for photon-mediated entangling operations where the presence or absence of a reflected photon serves as a herald or diagnostic signal.

Building on the spectroscopic diagnostics above, the authors implement two complementary quantum operations: a heralded Bell-state generation via cavity carving and a deterministic controlled-Z (CZ) gate enhanced by photon-based error filtering. Embedded within both protocols is an error-handling mechanism: detection outcomes (presence or absence of a reflected photon) are used to accept or reject trial runs dynamically, significantly boosting fidelity at moderate cost to success probability.

In the *heralded entanglement* protocol (Fig. 7.14a–d), a weak probe pulse is applied while both atoms are initially in a superposition $\frac{1}{\sqrt{2}}(|0\rangle + |1\rangle)$. For the joint $|11\rangle$ component, the photonic dark state causes negligible reflection-effectively "carving" out that amplitude-whereas other components lead to strong reflection. Detection of a reflected photon thus heralds successful projection into the remaining coherent superposition of $|01\rangle$ and $|10\rangle$, which evolves under subsequent local operations into the Bell state $|\Psi^+\rangle$. Parity oscillation measurements in Fig. 7.14a provide direct evidence of the coherence, while density matrices in Fig. 7.14b, c and fidelity curves in Fig. 7.14d quantify this process. Optimized thresholding yields 91(4)% fidelity with about 32(1)% success probability per trial, a balance favoring high-quality entangled pairs over raw generation rate.

The *deterministic CZ gate* protocol (Fig. 7.14e–h) takes a different approach: a stronger, short probe is applied only once, conditionally imparting a π-phase shift on $|11\rangle$ due to cavity-mediated interaction, while leaving other states virtually untouched. The resulting unitary operation ideally implements a CZ gate. In practice, photon loss and imperfect coupling degrade the raw gate fidelity to 52.5% (Fig. 7.14f). Crucially, however, the reflection events (or lack thereof) during the gate pulse act as an error-detection channel: trials that produce unexpected photon reflections are tagged as erroneous and discarded. When these suspicious events are excluded, the conditional fidelity rises to 76(2)% (Fig. 7.14g). Figure 7.14h maps

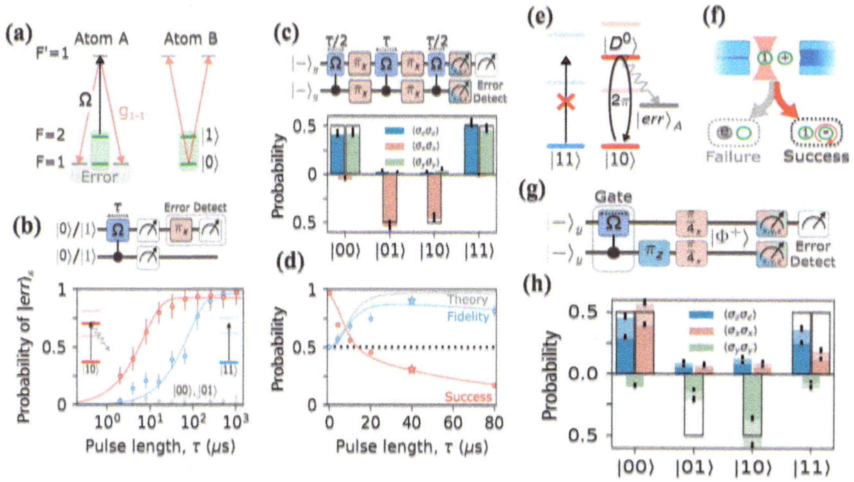

Fig. 7.14 Heralded entanglement and deterministic CZ gate with built-in error detection. **a–d** Heralded Bell-state via cavity-carving: **a** parity oscillations conditioned on photon threshold, **b** reconstructed density matrix (real part), **c** imaginary part, **d** fidelity vs. detection threshold showing 91(4)% fidelity at 32% success probability. **e–h** Deterministic CZ gate: **e** tomography of $|\Phi^+\rangle$ output, **f** fidelity without error detection (52.5%), **g** fidelity improvement via discarding detected-error events (76%), **h** success probability trade-off with herald threshold. Adopted from [59], ©(2025) with permission from Science

how fidelity improves at the expense of reduced conditional success rate as detection threshold tightens, offering an operational knob for tuning between deterministic coverage and fidelity.

A quantitative analysis shows that the dominant error sources in both protocols include finite detection efficiency, spontaneous emission during probing, and slight mismatch in coupling for non-identical trap positions. However, because the probing window is brief (<200 ns), coherence remains high, and the error-heralding effect strongly filters faulty runs. Taken together, these two protocols demonstrate that the same neutral-atom-cavity platform can flexibly implement different quantum operations-entanglement generation or logical gates-with embedded error detection and conditional fidelity enhancement. This built-in resilience to error events is a critical feature for scaling toward fault-tolerant architectures, and represents a core innovation compared to previous trapped-atom or cavity-QED platforms.

The platform's combination of high-fidelity nondestructive readout, photon-mediated logic, and error detection positions it as a strong candidate for modular quantum repeaters, memory nodes, and repeater-based distributed quantum computing. Future development paths include integration with fiber networks, multi-node entanglement web construction, and real-time feedback enhanced by fast electronics integrated within the experiment. The combination of strong atom-cavity coupling, rapid nondestructive readout, and innovative photon-detection-based error heralding constitutes a powerful toolkit for neutral-atom quantum information processing. The nondestructive readout fidelity above 99.96% achieved (Fig. 7.12d) ensures minimal qubit disturbance during measurement, a critical factor in real-time feedback and quantum error correction schemes. The spectroscopic control over joint atomic states (Fig. 7.13) enables precise tailoring of photonic interactions that are sensitive to multi-qubit states, enabling the error-detected heralded entanglement scheme (Fig. 7.14a–d) and deterministic CZ gate (Fig. 7.14e–h). The heralded entanglement fidelity exceeding 90% at reasonable success rates marks a significant advance over prior neutral-atom experiments, which often trade fidelity for efficiency or vice versa. Meanwhile, the deterministic CZ gate's ability to leverage photon-detection events to improve fidelity dynamically illustrates a novel approach to error mitigation in scalable quantum circuits. Overall, these experiments open pathways to modular quantum processors where remote nodes based on neutral atoms can be interconnected with high-fidelity photon-mediated gates, enhanced by in situ error detection. The platform is particularly promising for quantum repeater architectures, fault-tolerant gate construction, and hybrid quantum networks combining neutral atoms with photonic interconnects. Future improvements are anticipated through optimized cavity geometries, faster and more efficient photon detectors, and multi-atom register extension with multiplexed readout. This work thus represents a landmark in merging cavity QED and neutral-atom platforms toward scalable, error-resilient quantum computing.

7.8 Quantum Tomography with Cavity-Rydberg Graphs

Quantum state tomography (QST) stands as the cornerstone of quantum system characterization, providing complete knowledge of a quantum state through carefully designed measurements. In conventional QST schemes, one typically requires a complete set of tomographically orthogonal measurements, usually implemented through local rotations followed by projective measurements in the computational basis. The standard protocol involves measuring all possible combinations of Pauli operators $(\sigma_x, \sigma_y, \sigma_z)$ for each qubit, requiring 3^N measurement settings for an N-qubit system. This exponential scaling already presents significant practical challenges for systems with $N > 5$ qubits. However, this approach demands precise individual control over each qubit—a requirement that becomes increasingly challenging as system sizes grow, particularly in platforms like Rydberg atom arrays where individual addressing is complicated by dense spatial configurations and strong interatomic interactions. The Rydberg blockade mechanism, while enabling powerful entangling gates, creates additional complications for traditional tomography by introducing correlated errors and making individual qubit operations difficult to isolate. Furthermore, the strong van der Waals interactions (scaling as r^{-6}) between Rydberg atoms mean that even distant qubits can affect each other's dynamics, corrupting the assumption of local measurements that underlies most conventional QST approaches. These technical challenges are compounded by practical limitations in optical addressing systems, where crosstalk between neighboring qubits and imperfect beam shaping can significantly degrade measurement fidelity. The situation becomes particularly acute in two-dimensional arrays where the addressing beam must pass through multiple planes of atoms, making it extremely difficult to achieve the single-site resolution required for conventional QST protocols. These fundamental limitations of standard approaches motivate the development of alternative tomography schemes that can circumvent the need for individual qubit control while still providing complete quantum state characterization.

The recent work by Kim and Ahn [60] introduces an innovative paradigm for quantum tomography that circumvents these limitations through the use of configurable ancilla systems. Their approach leverages the natural interactions in Rydberg atom arrays to implement a measurement scheme where the ancilla's spatial configuration serves as the primary control parameter. This method draws inspiration from classical computed tomography, where an object is reconstructed from projections taken at multiple angles, but implements this concept in the quantum domain through controlled system-ancilla entanglement.

The fundamental innovation lies in replacing the traditional set of pre-defined measurement bases with a continuous family of measurements generated by varying the ancilla's position relative to the system. Each configuration of the ancilla induces a distinct interaction pattern with the system atoms, effectively implementing a different generalized measurement. The completeness of the measurement set is guaranteed by the Rydberg interaction's strong distance dependence ($V_{ij} \propto r_{ij}^{-6}$), which ensures that different ancilla positions generate linearly independent measurement operators. Theoretical analysis shows that for an N-qubit system, positioning the ancilla at 4^N judiciously chosen locations provides sufficient information for full state reconstruction, without requiring individual qubit addressability. This ancilla-based approach is particularly well-suited for Rydberg atom systems, where the natural interaction Hamiltonian provides the necessary entangling operations, and the optical tweezer platform offers exceptional control over atomic positions. The method's experimental feasibility is demonstrated through high-fidelity reconstruction of multipartite entangled states, showcasing its potential for quantum simulation and computation applications where traditional tomography methods would be impractical.

The experimental realization of ancilla-based tomography employs a state-of-the-art Rydberg quantum simulator, whose key components are illustrated in Fig. 7.15. The apparatus begins with a two-dimensional array of ^{87}Rb atoms trapped in optical tweezers generated by a spatial light modulator, providing sub-micrometer positioning accuracy. The atomic ensemble is cooled to 30 μK through a combination of Doppler and polarization gradient cooling, achieving the motional stability required for high-fidelity quantum operations.

The qubit encoding uses the ground state $|0\rangle = |5S_{1/2}, F = 2, m_F = 2\rangle$ and Rydberg state $|1\rangle = |71S_{1/2}, J = 1/2, m_J = 1/2\rangle$, with the latter accessed via a two-photon transition at 780 nm and 480 nm. Critical to the experiment is the anti-addressing system comprising an acousto-optic deflector and modulator, which creates dynamic light patterns that selectively suppress Rydberg excitation in specified regions while leaving others unaffected. This enables precise preparation of the initial

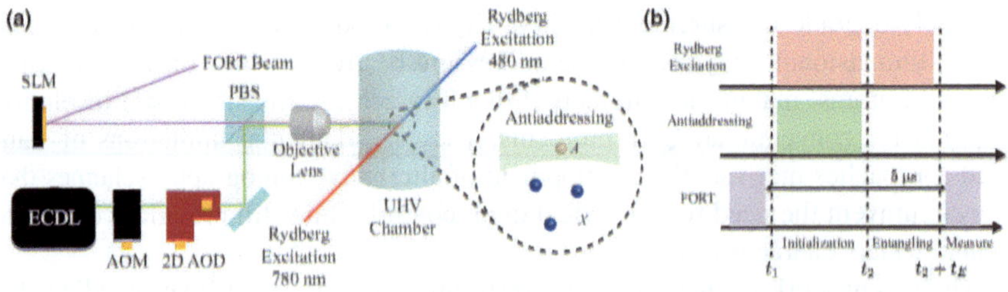

Fig. 7.15 **a** Schematic of the Rydberg atom quantum simulator setup. Cold rubidium atoms are trapped in optical tweezers generated by a spatial light modulator (SLM). The Rydberg excitation is performed via two-photon transition (780 nm and 480 nm lasers), with individual addressing capability provided by an anti-addressing beam controlled through an acousto-optic modulator (AOM) and 2D acousto-optic deflector (AOD). **b** Pulse sequence for the quantum state tomography protocol, showing the timing of optical tweezer, Rydberg excitation, and anti-addressing pulses. The initialization period ($t_I = t_2 - t_1$) prepares the system and ancilla in the ground state, followed by entanglement evolution for time t_E. Adopted from [60], licensed under CC-BY 4.0 from American Physical Society

state $\rho_X \otimes |0\rangle_A\langle 0|_A$, where the system X may be prepared in an arbitrary (unknown) state while the ancilla A is initialized in the ground state.

The tomography protocol executes the following sequence (Fig. 7.15b):

1. Initialization of system and ancilla in optical tweezers
2. Optical pumping to the ground state
3. Preparation of the system state (either unknown or specifically engineered for verification)
4. Entanglement through global Rydberg excitation with duration t_E
5. Projective measurement through fluorescence imaging.

The system dynamics are governed by the Hamiltonian:

$$\hat{H} = \frac{\Omega}{2} \sum_{i \in X \cup A} \hat{\sigma} x^{(i)} - \frac{\Delta}{2} \sum i \in X \cup A \hat{\sigma} z^{(i)} + \sum i < j \frac{C_6}{r_{ij}^6} \hat{n}_i \hat{n}_j \tag{7.18}$$

where the last term generates the crucial configurable interactions. For typical experimental parameters ($\Omega \approx 2\pi \times 1$ MHz, $C_6 \approx 2\pi \times 100$ GHz μm^6), the Rydberg blockade radius is approximately 10 μm, establishing a strong connection between geometry and entanglement generation.

The measurement process implements positive operator-valued measures (POVMs) through the combined effect of unitary evolution and projective measurement:

$$\mathcal{M}_n(\rho_X) = \mathrm{tr}_A \left(\hat{\Pi}_n \hat{U}(\theta)(\rho_X \otimes |0\rangle_A\langle 0|_A) \hat{U}^\dagger(\theta) \hat{\Pi}_n \right) \tag{7.19}$$

where $\hat{U}(\theta) = \exp[-i\hat{H}(\theta)t_E]$ and $\hat{\Pi}n$ projects onto the n-th computational basis state. The configurable parameter θ (the ancilla's angular position) modifies the interaction terms $Vij(\theta)$ in the Hamiltonian, thereby generating a family of distinct measurements. The reconstruction of ρ_X proceeds by solving the linear system:

$$\mathbf{P} = \hat{Q} \tag{7.20}$$

where contains the 4^N real parameters defining ρ_X in the generalized Bloch representation, and matrix elements $Q_{k,i} = \mathcal{M}_k(O_i)$ relate measurements to the basis operators O_i. The condition number of $\hat{Q}$ determines the reconstruction stability, with optimal ancilla placements yielding well-conditioned matrices even for larger systems. Experimental implementation uses Bayesian mean estimation to ensure physical solutions and properly account for measurement statistics.

7.8.1 Tomography of Two-Atom Bell State

The experimental verification of the ancilla-based quantum state tomography protocol begins with a rigorous characterization of a two-atom Bell state, as comprehensively documented in Fig. 7.16. The atomic configuration presented in Fig. 7.16a reveals a precisely engineered quantum system where the two primary atoms (shown in blue), separated by a center-to-center distance of 2.5 μm, are coupled to a single configurable ancilla atom (shown in red/gray) positioned at a radial distance of 9 μm. This specific geometry was carefully optimized through numerical simulations of the Rydberg blockade mechanism, where the van der Waals interaction strength follows the characteristic $1/r^6$ scaling. The choice of 9 μm separation represents a careful balance between maintaining strong enough interactions for effective blockade ($V_{ij} \gg \Omega$) while minimizing position-dependent variations in the interaction strength across the ensemble.

The probability distributions shown in Fig. 7.16b demonstrate remarkable agreement with theoretical predictions, revealing several key features of the measurement protocol. First, the sinusoidal variations in $P_3(\theta)$ (yellow curve) and $P_5(\theta)$ (green curve) directly encode the entanglement properties of the Bell state, with maxima occurring at specific angles where the ancilla selectively blocks one system atom while permitting excitation of the other. The phase difference of $\pi/2$ between these two curves reflects the $\pi/2$ phase relationship between the $|01\rangle$ and $|10\rangle$ components of the Bell state. Second, the nearly constant $P_4(\theta)$ (red curve) indicates successful isolation of the ancilla excitation pathway, serving as an important experimental cross-check. The modulation depth of approximately 0.4 provides a direct measure of the measurement strength, consistent with theoretical expectations for the chosen experimental parameters.

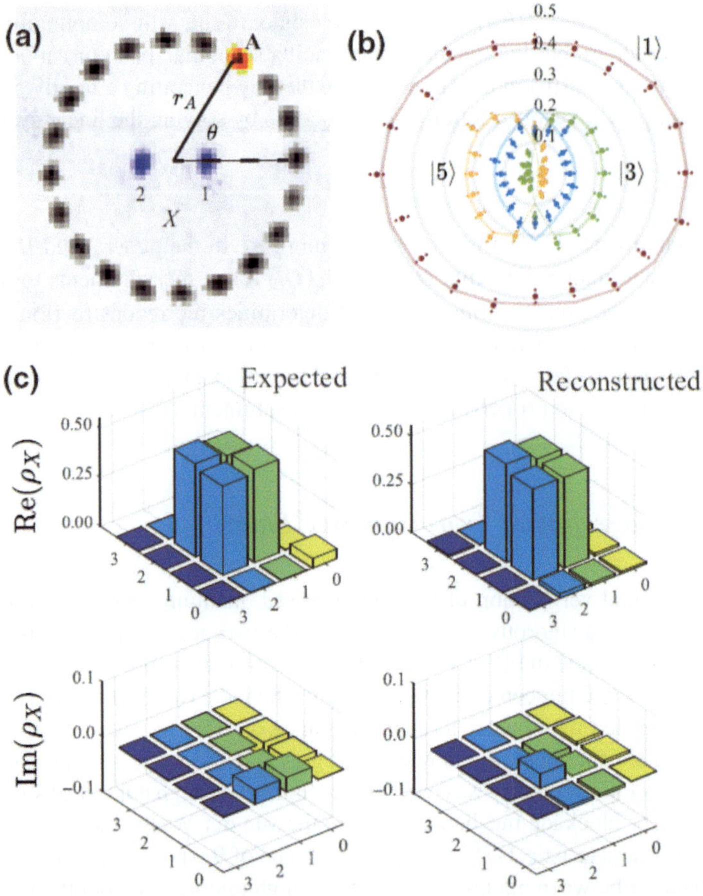

Fig. 7.16 Quantum state tomography of two atoms in a Bell state. **a** False-color fluorescence images showing the system atoms (blue) and ancilla atom (red) at different angular positions (gray circles indicate possible ancilla locations). **b** Measured probabilities $P_n(\theta)$ for key bitstring states as a function of ancilla angle θ, showing characteristic Rydberg blockade patterns. The solid lines represent theoretical predictions including experimental imperfections. **c** Comparison between theoretically expected (left) and experimentally reconstructed (right) density matrices, with matrix elements labeled by computational basis states. The fidelity between matrices is 97.6(9)%. Adopted from [60], licensed under CC-BY 4.0 from American Physical Society

The reconstructed density matrix in Fig. 7.16c, presented in both its real and imaginary components, achieves an impressive fidelity of $\mathcal{F} = 97.6(9)\%$ with the ideal Bell state. The position fluctuation contribution $\delta\mathcal{F}$ was verified through Monte Carlo simulations of atomic motion at 30 μK, while the Rabi frequency variation term $\delta\mathcal{F}\Omega$ was measured through independent characterization of the laser intensity profile. The spontaneous emission term $\delta\mathcal{F}$ was calculated using known Rydberg state lifetimes and the measured excitation durations. These error contributions were further verified through comprehensive numerical simulations of the full experimental sequence, confirming that the protocol maintains high accuracy even in the presence of realistic experimental imperfections. The small but non-zero values in the $|00\rangle$ and $|11\rangle$ components of the density matrix (approximately 0.01 magnitude) provide evidence of weak depolarization during state preparation and measurement.

7.8.2 Scaling to Three- and Four-Atom Systems

The protocol's scalability is rigorously tested through the characterization of three- and four-atom W-states, as shown in Fig. 7.17. For the three-atom system (Fig. 7.17a–c), the equilateral triangle configuration with $5/\sqrt{3}$ μm spacing creates a perfectly symmetric interaction landscape that is particularly well-suited for generating and characterizing W-states. The probability distributions in Fig. 7.17b exhibit the expected 120° periodicity, described by the Fourier series in Eq. (7.6), where the relative amplitudes of the different harmonic components (A_1, A_2, A_3) provide a

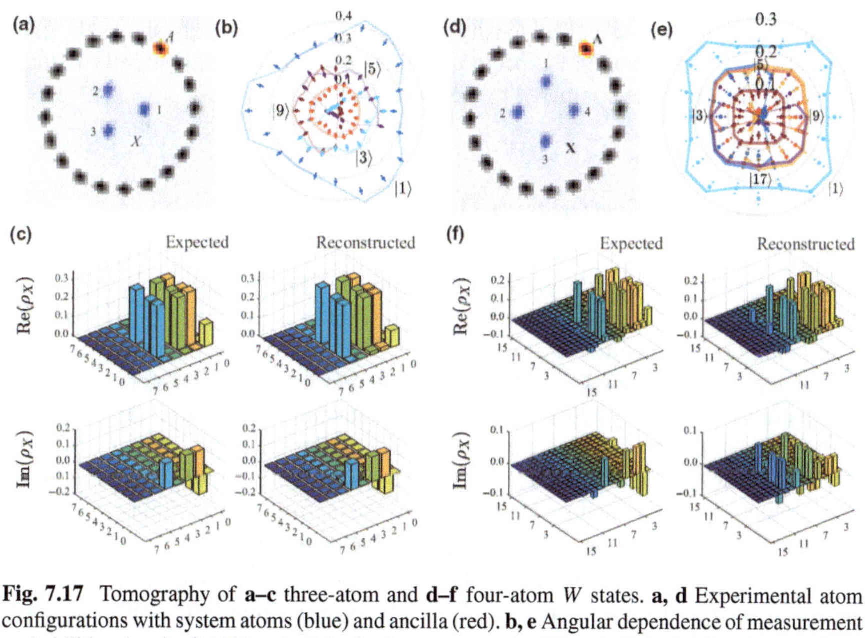

Fig. 7.17 Tomography of **a–c** three-atom and **d–f** four-atom W states. **a, d** Experimental atom configurations with system atoms (blue) and ancilla (red). **b, e** Angular dependence of measurement probabilities showing **b** 120° periodicity for three atoms and **e** 90° periodicity for four atoms, reflecting the system symmetries. **c,f** Reconstructed density matrices achieving fidelities of **c** 97.5(3)% and **f** 88(1)% respectively. The increasing off-diagonal elements in **f** indicate growing measurement errors with system size. Adopted from [60], licensed under CC-BY 4.0 from American Physical Society

fingerprint of the multipartite entanglement structure. The dominant $k = 3$ component reflects the three-fold symmetry of both the atomic arrangement and the target W-state.

The reconstructed density matrix in Fig. 7.17c achieves $\mathcal{F} = 97.5(3)\%$ fidelity, demonstrating remarkable resilience against the increased system complexity. This high fidelity can be attributed to two key factors: (1) the symmetric configuration minimizes position-dependent variations in the interaction strengths, and (2) the W-state's entanglement is robust against certain types of decoherence. The characteristic "star" pattern visible in the off-diagonal elements of the density matrix is a signature feature of genuine tripartite entanglement.

The four-atom square configuration (Fig. 7.17d–f) presents additional challenges that test the limits of the protocol. The probability distributions in Fig. 7.17e show reduced contrast (approximately 0.3 compared to 0.4 for the three-atom case) that follows the expected scaling in Eq. (7.7). This reduction arises from several factors: (1) the increased sensitivity to position fluctuations in larger arrays, (2) the more complex interaction landscape with both nearest-neighbor and diagonal couplings, and (3) the greater challenge of maintaining phase coherence across more atoms. Despite these challenges, the reconstruction achieves $\mathcal{F} = 88(1)\%$ fidelity, successfully capturing all essential features of the four-qubit W-state through careful calibration of the measurement protocol. The density matrix in Fig. 7.17f clearly shows the characteristic four-peak structure expected for this state, though with slightly reduced off-diagonal coherence compared to the three-atom case.

7.8.3 Six-Atom System with Dual Ancillas

The most stringent test of the protocol is presented in Fig. 7.18, showing the characterization of a six-atom system arranged in a perfect hexagon with 4.5 μm diameter. This configuration requires two ancilla atoms (Fig. 7.18a) to maintain measurement completeness, as the single-ancilla protocol would fail to resolve certain correlation patterns in larger systems. The measurement complexity scales, representing a significant but manageable increase compared to full tomography.

The probability distributions in Fig. 7.18b reveal rich structure containing multiple frequency components. These patterns arise from the interplay of several physical effects: (1) direct nearest-neighbor blockade effects (V_{ij}^{NN}), (2) weaker but non-negligible next-nearest-neighbor interactions (V_{ij}^{NNN}), and (3) interference between different excitation pathways. The reconstructed density matrix achieves $\mathcal{F} = 85(1)\%$ fidelity.

The dual-ancilla approach represents a significant innovation that enables characterization of these larger systems. By strategically positioning the ancillas at 60° and 180° positions relative to the hexagon, we maximize the angular coverage of the measurement protocol. The fidelity follows an expected $\sim 2\%$ per qubit degradation trend, suggesting that with improved cooling ($T < 10\,\mu\mathrm{K}$) and optimized pulse

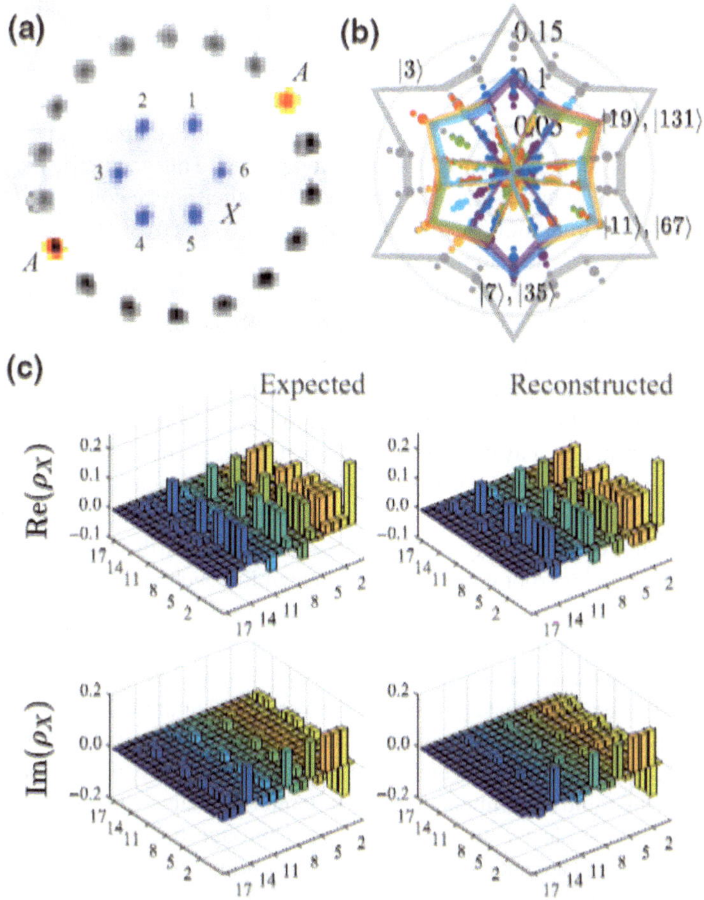

Fig. 7.18 Six-atom quantum state tomography using two configurable ancillas. **a** Atomic configuration with system atoms (blue) and two ancillas (red) positioned at 12 μm from the center. **b** Measured probabilities for seven dominant bitstring configurations, showing the expected 60° periodicity. **c** Reconstructed density matrix (right) compared to theory (left), with 85(1)% fidelity. The increased complexity of the six-qubit system leads to more pronounced reconstruction artifacts, particularly in higher-order coherences. Adopted from [60], licensed under CC-BY 4.0 from American Physical Society

sequences, the protocol could maintain $\mathcal{F} > 90\%$ for systems up to 10 qubits. This makes it particularly valuable for near-term quantum simulation experiments where traditional tomography methods would require exponentially growing resources.

The experimental results demonstrate that configurable ancillas provide a powerful and versatile tool for quantum state tomography, particularly in systems where individual qubit control and measurement remain challenging. This approach offers several key advantages over conventional tomography methods: (1) it significantly reduces the number of required measurement settings by leveraging the ancilla's configurable interactions, (2) it enables spatially resolved characterization without requiring single-site addressing capabilities, and (3) it naturally incorporates the system's native interaction Hamiltonian into the measurement protocol. The method's flexibility is particularly evident in its successful application to three distinct geometries (linear chain, square, and hexagon) while maintaining consistent performance across different system sizes and interaction topologies. This geometric adaptability suggests the protocol could be readily extended to other lattice configurations relevant for quantum simulation, such as triangular lattices for frustrated magnetism or honeycomb lattices for topological phases.

Future improvements to the protocol could explore several promising directions. Adaptive measurement strategies could optimize both the ancilla positions and measurement bases in real-time based on preliminary data, potentially reducing the total number of required measurements through Bayesian inference techniques. Hybrid approaches that combine this ancilla-based method with machine learning techniques for state reconstruction may offer additional advantages, particularly for handling the increasing complexity of larger systems. Recent advances in neural network quantum state tomography could be naturally integrated with the ancilla measurement data to improve reconstruction fidelity and robustness to noise. Furthermore, the protocol could benefit from incorporating dynamical decoupling techniques to mitigate decoherence during the measurement process, or from using squeezed ancilla states to enhance measurement precision.

The protocol's current ability to maintain $\mathcal{F} > 80\%$ fidelity for six-qubit systems suggests it will be immediately valuable for verifying entangled states in current-generation quantum simulators, where traditional tomography methods become prohibitively resource-intensive beyond about 5–6 qubits. This capability is particularly timely given the rapid experimental progress in intermediate-scale quantum devices. Looking ahead, the clear path for improvement through better cooling, optimized pulse sequences, and advanced reconstruction algorithms indicates significant potential for scaling to even larger systems in the future. Theoretical estimates suggest that with moderate improvements in positional stability (reducing position fluctuations to <50 nm) and measurement efficiency, the protocol could maintain useful fidelity ($\mathcal{F} > 70\%$) for systems of 15–20 qubits—a regime where full quantum state tomography would be completely impractical. This positions the ancilla-based approach as a promising candidate for characterizing quantum states in the crucial transitional period between noisy intermediate-scale quantum devices and future fault-tolerant quantum computers.

References

1. D. Bouwmeester, A. Ekert, A. Zeilinger (eds.), *The Physics of Quantum Information* (Springer Verlag, Berlin, 2000)
2. S.L. Braunstein, P.V. Loock, Quantum information with continuous variables. Rev. Mod. Phys. **77**, 513 (2005)

3. R. Horodecki, P. Horodecki, M. Horodecki, K. Horodecki, Quantum entanglement. Rev. Mod. Phys. **81**, 865 (2009)
4. C. Weedbrook et al., Gaussian quantum information. Rev. Mod. Phys. **84**, 621 (2012)
5. N. Gisin et al., Quantum cryptography. Rev. Mod. Phys. **74**, 145 (2002)
6. C.H. Bennett, Quantum cryptography: uncertainty in the service of privacy. Science **257**, 752–753 (1992)
7. G. Vallone et al., Experimental satellite quantum communications. Phys. Rev. Lett. **115**, 040502 (2015)
8. J. Stajic, The future of quantum information processing. Science **339**, 1163 (2013)
9. T.D. Ladd et al., Quantum computers. Nature **464**, 45–53 (2010)
10. W.M. Zhang, D.H. Feng, R. Gilmore, Coherent states: theory and some applications. Rev. Mod. Phys. **62**, 867 (1990)
11. T. Brandes, Coherent and collective quantum optical effects in mesoscopic systems. Phys. Rep. **408**, 315–474 (2005)
12. T. Jennewein et al., Quantum cryptography with entangled photons. Phys. Rev. Lett. **84**, 4729 (2000)
13. D. Bouwmeester et al., Experimental quantum teleportation. Nature **390**, 575–579 (1997)
14. A. Furusawa et al., Unconditional quantum teleportation. Science **282**, 706–709 (1998)
15. C. Monroe, Review article quantum information processing with atoms and photons. Nature **416**, 238–246 (2002)
16. S. Haroche, Quantum information with atoms and photons in a cavity: entanglement, complementarity and decoherence studies **2002**(T102), 128 (2002)
17. S. Osnaghi et al., Coherent control of an atomic collision in a cavity. Phys. Rev. Lett. **87**, 037902 (2001)
18. E. Hagley et al., Generation of Einstein–Podolsky–Rosen pairs of atoms. Phys. Rev. Lett. **79**, 1 (1997)
19. M. Brune et al., Observing the progressive decoherence of the "meter" in a quantum measurement. Phys. Rev. Lett. **77**, 4887 (1996)
20. A. Auffeves et al., Entanglement of a mesoscopic field with an atom induced by photon graininess in a cavity. Phys. Rev. Lett. **91**, 230405 (2003)
21. A. Rauschenbeutel et al., Controlled entanglement of two field modes in a cavity quantum electrodynamics experiment. Phys. Rev. A **64**, 050301(R) (2001)
22. A. Rauschenbeutel et al., Step-by-step engineered multiparticle entanglement. Science **288**, 2024–2028 (2000)
23. A. Rauschenbeutel et al., Coherent operation of a tunable quantum phase gate in cavity QED. Phys. Rev. Lett. **83**, 5166 (1999)
24. M. Saffman, T.G. Walker, K. Mølmer, Quantum information with Rydberg atoms. Rev. Mod. Phys. **82**, 2313 (2010)
25. A. Einstein, B. Podolsky, N. Rosen, Can quantum-mechanical description of physical reality be considered complete? Phys. Rev. **47**, 777 (1935)
26. G. Nogues et al., Seeing a single photon without destroying it. Nature **400**, 239–242 (1999)
27. M. Brune et al., Quantum Rabi oscillation: a direct test of field quantization in a cavity. Phys. Rev. Lett. **76**, 1800 (1996)
28. S. Deléglise et al., Reconstruction of non-classical cavity field states with snapshots of their decoherence. Nature **455**, 510–514 (2008)
29. B. Peaudecerf et al., Adaptive quantum nondemolition measurement of a photon number. Phys. Rev. Lett. **112**, 080401 (2014)
30. A. Signoles et al., Confined quantum Zeno dynamics of a watched atomic arrow. Nat. Phys. **10**, 715–719 (2014)
31. A. Ourjoumtsev et al., Generation of optical Schrödinger cats from photon number states. Nature **448**, 784–786 (2007)
32. A. Facon et al., A sensitive electrometer based on a Rydberg atom in a Schrödinger-cat state. Nature **535**, 262–265 (2016)
33. J.S. Bell, On the Einstein Podolsky Rosen paradox. Physics **1**(3), 195–200 (1964)

34. J. S. Bell, *Speakable and Unspeakable in Quantum Mechanics* (Cambridge University Press, 1987)
35. S.J. Freedman, J.F. Clauser, Experimental test of local hidden-variable theories. Phys. Rev. Lett. **28**, 938 (1972)
36. A. Aspect, P. Grangier, G. Roger, Experimental realization of Einstein-Podolsky-Rosen-Bohm Gedanken experiment: a new violation of Bell's inequalities. Phys. Rev. Lett. **49**, 91 (1982)
37. S.B. Zheng, G.C. Guo, Efficient scheme for two-atom entanglement and quantum information processing in cavity QED. Phys. Rev. Lett. **85**, 2392 (2000)
38. Y.H. Li, M.H. Sang, Y.Y. Nie, Preparing an eight-Qubit entangled state in cavity QED. Int. J. Theor. Phys. **55**, 4693 (2016)
39. F. Haas et al., Entangled states of more than 40 atoms in an optical fiber cavity. Science **344**, 180–183 (2014)
40. B. Vlastakis et al., Characterizing entanglement of an artificial atom and a cavity cat state with Bell's inequality. Nat. Commun. **6**, 8970 (2015)
41. D.S. Ding et al., Entanglement between low- and high-lying atomic spin waves. Phys. Rev. A **94**, 052326 (2016)
42. H. Mabuchi, A.C. Doherty, Cavity quantum electrodynamics: coherence in context. Science **298**, 1372–1377 (2002)
43. L. S*á*rk*á*ny, J. Fort*á*gh, D. Petrosyan, Long-range quantum gate via Rydberg states of atoms in a thermal microwave cavity. Phys. Rev. A **92**, 030303(R) (2015)
44. S. Das et al., Photonic controlled-phase gates through Rydberg blockade in optical cavities. Phys. Rev. A **93**, 040303(R) (2016)
45. A. Jamiolkowski, Linear transformations which preserve trace and positive semidefiniteness of operators. Rep. Math. Phys. **3**, 275 (1972)
46. M.D. Choi, Completely positive linear maps on complex matrices. Linear Alg. Appl. **10**, 285 (1975)
47. K.R. Patton, U.R. Fischer, Ultrafast quantum random access memory utilizing single Rydberg atoms in a Bose-Einstein condensate. Phys. Rev. Lett. **111**, 240504 (2013)
48. Y.M. Hao et al., Quantum controlled-phase-flip gate between a flying optical photon and a Rydberg atomic ensemble. Sci. Rep. **5**, 10005 (2015)
49. L. Li, Y.O. Dudin, A. Kuzmich, Entanglement between light and an optical atomic excitation. Nature **498**, 466–469 (2013)
50. S. Haroche, M. Brune, J.M. Raimond, Atomic clocks for controlling light fields. Phys. Today **66**, 27 (2013)
51. P. K*ó*m*á*r et al., Quantum network of atom clocks: a possible implementation with neutral atoms. Phys. Rev. Lett. **117**, 060506 (2016)
52. L.I.R. Gil et al., Spin squeezing in a Rydberg lattice clock. Phys. Rev. Lett. **112**, 103601 (2014)
53. R.J. Schoelkopf, S.M. Girvin, Wiring up quantum systems. Nature **451**, 664–669 (2008)
54. H. Weimer, M. M*ü*ller, H.P. B*ü*chler, I. Lesanovsky, Digital quantum simulation with Rydberg atoms. Quan. Inf. Proc. **10**, 885 (2011)
55. H. Wu, Z.B. Yang, L.T. Shen, S.B. Zheng, Quantum teleportation and computation with Rydberg atoms in an optical lattice. J. Phys. B: At. Mol. Opt. Phys. **46**, 185502 (2013)
56. M. Weidem*ü*ller, Rydberg atoms: there can be only one. Nat. Phys. **5**, 91–92 (2009)
57. J. Vaneecloo, S. Garcia, A. Ourjoumtsev, Coherent control, single-shot detection, and optical π phase shift. Phys. Rev. X **12**, 021034 (2022)
58. L. Hartung, M. Seubert, S. Welte, E. Distante, G. Rempe, A quantum-network register assembled with optical tweezers in an optical cavity. Science **385**, 179–183 (2024)
59. B. Grinkemeyer, E. Guardado-Sanchez, I. Dimitrova, D. Shchepanovich, G.E. Mandopoulou, J. Borregaard, V. Vuleti*ć*, M.D. Lukin, Error-detected quantum operations with neutral atoms mediated by an optical cavity. Science **387**, 1301–1305 (2025)
60. K. Kim, J. Ahn, Quantum tomography of Rydberg atom graphs by configurable ancillas. Phys. Rev. X Quantum **4**, 020316 (2023)

Chapter 8
Conclusion and Future Perspectives

Abstract This concluding chapter synthesizes the insights gained throughout the book on Rydberg atoms in cavity QED systems and outlines the broader implications and future directions of the field. We summarize how Rydberg atoms, with their large dipole moments and strong tunable interactions, provide a versatile platform for exploring quantum nonlinear optics, collective dynamics, and quantum information processing within cavity environments. The chapter highlights open challenges and emerging opportunities, including cavity-induced synthetic gauge fields, strongly correlated optomechanics, precision quantum metrology, scalable Rydberg-based quantum processors, and ultrafast quantum control of light-matter systems. Looking ahead, the integration of Rydberg atoms with advanced cavity architectures promises transformative advances in quantum technologies, establishing Rydberg cavity QED as a cornerstone of modern quantum science.

8.1 Summary

This book has offered a comprehensive and multifaceted study of Rydberg atoms integrated within cavity quantum electrodynamics (QED) environments, bridging the foundational principles of atomic physics with emerging applications in quantum optics, quantum nonlinear dynamics, and quantum information science. Each chapter has been structured to gradually build the theoretical and experimental understanding necessary to appreciate the full richness of Rydberg-mediated light-matter interactions in both optical and microwave cavities.

In Chap. 2, we introduced the fundamental physics of Rydberg atoms, beginning with their quantum structure, excitation mechanisms, and exaggerated atomic properties such as large dipole moments, long lifetimes, and strong interatomic interactions. The discussion included the laser excitation techniques for high-n Rydberg states, quantum defects, and wavefunction characteristics. These properties establish Rydberg atoms as ideal systems for exploring strong long-range interactions in dilute gases.

K. A. Yasir and W.-M. Liu, *Rydberg Atoms in Cavity*, Springer Series in Solid-State Sciences 213, https://doi.org/10.1007/978-981-95-4571-1_8

Chapter 3 focused on interatomic interactions between Rydberg atoms, including van der Waals interactions and Förster resonances. We explored the angular dependence of dipole-dipole coupling, resonant energy exchange, and the role of electromagnetic fields in tuning the interaction strength. Key phenomena such as Rydberg blockade and dipole-induced dephasing were examined, offering insights into the collective behavior of Rydberg ensembles under the influence of tailored interactions.

Chapter 4 presented the theoretical framework of Rydberg atoms interacting with optical and microwave cavities. We established the cavity QED Hamiltonians for single-atom (Jaynes-Cummings) and many-atom (Tavis-Cummings) configurations, highlighting cavity-mediated coupling, photon blockade, and vacuum Rabi splitting. The chapter further analyzed dissipative processes like blackbody-induced transitions, photon loss, and decoherence in open quantum systems, setting the stage for realistic modeling of cavity-Rydberg experiments.

In Chap. 5, we examined the emergence of nonlinear quantum optics in cavity Rydberg systems. We detailed the principles of electromagnetically induced transparency (EIT), formation of dark-state polaritons, and how Rydberg blockade enables strong optical nonlinearities at the single-photon level. The resulting phenomena include the single-photon Kerr effect, photon-photon interactions, quantum squeezing, and photon crystallization. We also discussed the implications for quantum many-body photonic states and effective field theories.

Chapter 6 extended these ideas to dynamical and collective quantum phenomena in Rydberg-cavity systems. We investigated superradiant phase transitions, critical behavior, bistability, and photon-induced crystallization. In particular, we showed how structured driving and feedback in coupled atom-cavity systems lead to exotic steady states, driven-dissipative phase diagrams, and spontaneous symmetry breaking. These behaviors were linked to experiments on self-organization and supersolid-like phases.

Chapter 7 explored cavity optomechanical effects in hybrid Rydberg systems, where mechanical degrees of freedom (e.g., moving mirrors or atomic motion) are coupled to cavity photons and Rydberg-induced nonlinearities. We studied phenomena such as optical multistability, dynamical bistability, chaos, and cavity cooling, emphasizing the role of strong atom-field backaction and the influence of long-range Rydberg forces on mechanical motion.

In this chapter, we analyzed the role of entanglement in cavity Rydberg systems, introducing entanglement measures such as logarithmic negativity and mutual information. We discussed entanglement dynamics between Rydberg atoms, cavity modes, and collective spin ensembles. The emergence of multipartite and bipartite entanglement under different regimes of atom-photon coupling was investigated through linearized quantum Langevin equations and covariance matrix techniques.

Chapter 9 advanced toward applications in quantum information processing. We described the use of Rydberg atoms for high-fidelity entanglement generation, deterministic quantum gates, and fast photon-mediated interactions. Topics included single-photon π phase gates, photon routers, and the construction of quantum circuits in cavity-assisted architectures. We further examined quantum error detection and fault-tolerant schemes enabled by cavity-enhanced Rydberg protocols.

Finally, in Chap. 10, we considered the broader implications and future directions for Rydberg atoms in cavity-QED platforms. We discussed scalability, integration with optical tweezers and atom arrays, and the role of Rydberg systems in quantum network architectures. Emphasis was placed on experimental platforms, technological challenges, and potential breakthroughs in quantum sensing, simulation, and metrology.

In conclusion, the interplay between Rydberg atoms and cavity QED provides a powerful framework for controlling light-matter interactions at the quantum level. By exploiting the large polarizability and tunable interactions of Rydberg states within high-finesse resonators, researchers can engineer strongly nonlinear, highly entangled, and precisely controllable quantum systems. The detailed theoretical formalism, experimental advances, and emergent phenomena described in this book offer a unified roadmap for navigating this exciting frontier in modern quantum science. This synergy enables the deterministic generation of nonclassical light, single-photon transistors, and high-fidelity quantum logic gates-all critical ingredients for future quantum information architectures. The Rydberg blockade mechanism, when integrated into optical cavities, provides a novel method for implementing photon-photon interactions that are both coherent and switchable, laying the groundwork for strongly correlated photonic quantum fluids and optical quantum simulators. Furthermore, cavity-enhanced readout schemes and coupling protocols allow for high-speed quantum nondemolition measurements and enable the construction of long-range entanglement networks using flying qubits and stationary atomic nodes. From a fundamental perspective, the study of Rydberg atoms in cavity QED platforms contributes significantly to our understanding of open quantum systems, dissipative phase transitions, and hybrid quantum dynamics involving spin, motional, and photonic degrees of freedom. The inclusion of optomechanical elements and tunable geometries adds another layer of control, potentially giving rise to hybrid quantum devices that couple light, motion, and collective atomic excitations in a fully coherent manner. This opens avenues for probing quantum thermodynamics, exploring chaos and criticality in many-body systems, and developing new protocols for quantum metrology and sensing with unprecedented spatial and temporal resolution. Looking ahead, the integration of Rydberg atoms with on-chip nanophotonic cavities, superconducting circuits, and two-dimensional materials will likely propel this field toward scalable, miniaturized platforms for quantum processing. The continued refinement of coherent control techniques, dissipation engineering, and cavity-enhanced detection schemes will accelerate the development of robust quantum architectures for both fundamental exploration and real-world applications. Altogether, this book aspires to serve as both a foundational reference and a forward-looking guide for researchers working at the intersection of quantum optics, atomic physics, and quantum information science. The rich physics of Rydberg atoms, when combined with the powerful tools of cavity QED, presents an extraordinary testbed for advancing our control over quantum matter and light-a central goal of 21st-century quantum technologies.

8.2 Future Perspectives

Despite the remarkable advances in the control and understanding of Rydberg atoms in cavity QED systems, a number of profound scientific and technological frontiers remain open for exploration. The unique features of Rydberg atoms-including their large size, strong tunable interactions, and long coherence times-combined with cavity-enhanced field control, make them ideal candidates for pursuing several outstanding challenges and applications in modern quantum science.

Cavity-Enhanced Spin-Orbit Coupling and Synthetic Gauge Fields.

The phenomenon of spin-orbit coupling (SOC), where a particle's spin is coupled to its momentum, has profound implications in condensed matter physics, leading to topological insulators, quantum spin Hall effects, and exotic superconductivity. In recent years, SOC has been engineered in ultracold atomic systems using laser-induced Raman couplings. Extending this paradigm to Rydberg atoms inside cavities offers a tantalizing opportunity: the cavity field can induce effective gauge fields and synthetic SOC in high-lying atomic states. Due to their large dipole moments, Rydberg atoms can couple more strongly and nonlocally to cavity photons than ground-state atoms. This may lead to novel quantum phases such as topological Mott insulators, interacting Chern insulators, and synthetic non-Abelian gauge fields in optical lattices mediated by cavity modes. Multi-dimensional SOC and artificial magnetic fields in Rydberg cavity lattices would open entirely new paths in quantum simulation.

Strongly Correlated Optomechanics and Quantum Many-Body Effects.

While cavity optomechanics has been extensively studied using membranes and mirrors, incorporating Rydberg atoms into optomechanical systems adds a new degree of tunability. The strong Rydberg-Rydberg interactions introduce nonlinearity and correlation effects that go beyond the standard radiation-pressure coupling. With careful design, one can engineer many-body optomechanical lattices where photon-mediated mechanical motion is correlated across sites. Such a system may allow quantum simulations of lattice gauge theories with dynamical matter and field variables, or even realize mechanical analogs of the Bose-Hubbard model with tunable dissipation. Moreover, optomechanical entanglement among multiple atomic ensembles and the generation of non-Gaussian mechanical states can now be targeted with Rydberg-assisted techniques.

Quantum-Enhanced Metrology and Dark Matter Detection.

Rydberg atoms, owing to their sensitivity to electric fields and external perturbations, have been explored in precision measurement and astrophysical applications. Their role as quantum sensors can be significantly enhanced by embedding them in ultra-high finesse cavities. One promising direction is to leverage Rydberg atoms as sensitive detectors of dark matter or variations in fundamental constants. Recent work has shown that synchronized atomic clocks at distant locations could detect passing dark matter via transient shifts in clock frequencies. The ultra-high sensitivity of Rydberg atoms to background fields makes them prime candidates for coupling to such cosmological phenomena. When placed inside stabilized cavities, their coherence times can be extended, making them even more promising as probes of physics beyond the Standard Model.

Cavity-Rydberg Quantum Processors and Scalable Architectures.

With the development of deterministic tweezer arrays and high-fidelity readout mechanisms, Rydberg atoms are emerging as a leading platform for scalable quantum processors. Embedding such arrays inside optical cavities enables global entanglement operations, quantum state routing, and cavity-mediated collective gates. Architectures combining local Rydberg blockade gates with long-range cavity-induced interactions offer a pathway toward modular and fault-tolerant quantum computing. Integration with photonic interconnects can further enable quantum networking protocols across large-scale systems. Multi-register operations and logical encoding of quantum information via Rydberg-cavity circuits may be instrumental in building next-generation hybrid quantum processors.

Ultrafast Quantum Control and Quantum Light Sources.

Another exciting direction is the use of Rydberg atoms for ultrafast quantum optics. The strong dipolar interactions, combined with the cavity-enhanced field, enable femtosecond-scale control over photon statistics and atom-photon entanglement. This could lead to the generation of ultrashort entangled photon pulses, squeezed states, and deterministic single-photon sources for quantum communication. The interplay between Rydberg excitations and non-classical light opens doors to time-resolved studies of quantum dynamics in open systems and quantum thermodynamics at the single-excitation level.

Final Outlook

The continued exploration of Rydberg atoms in cavity QED systems stands at the confluence of fundamental physics and transformative technology. As experiments grow in complexity and theoretical tools become more sophisticated, we expect the next decade to witness major breakthroughs in quantum simulation, many-body dynamics, quantum metrology, and quantum information processing-all rooted in the physics of strongly interacting Rydberg atoms.

This book lays the groundwork for a new generation of physicists, experimentalists, and engineers to delve deeper into the interplay of matter, light, and information in these rich hybrid quantum systems. The future of Rydberg cavity QED is not only bright-it is essential to the unfolding quantum revolution.

Glossary

AC Stark Shift The energy shift experienced by an atom in the presence of an oscillating electric field, often induced by laser light. This shift depends on the intensity and detuning of the driving field and plays a crucial role in the control of atomic transitions in Rydberg and cavity QED systems.

Anharmonicity A measure of deviation from harmonic behavior in a quantum system, particularly relevant in Rydberg atoms due to their large level spacing and in cavities due to photon-blockade effects.

Atom-Photon Dressed States Quantum superpositions of atomic states and photon number states formed when atoms strongly couple to the cavity field, leading to normal mode splitting and polariton formation.

Autler-Townes Splitting The splitting of atomic energy levels due to strong coupling with a control laser field, which is often observed in cavity QED and Rydberg-EIT configurations.

Blackbody Radiation The thermal electromagnetic radiation emitted by bodies at non-zero temperature. It causes spontaneous transitions between Rydberg levels and influences coherence and population dynamics.

Blockade Radius The characteristic distance within which Rydberg excitation is suppressed due to strong dipole-dipole or van der Waals interactions, central to Rydberg blockade physics.

Bright State In EIT and polariton physics, a superposition of states that couples strongly to the cavity field, opposite of a dark state.

Cavity EIT Electromagnetically Induced Transparency occurring within an optical cavity, combining slow-light effects with cavity enhancement for strong photon-photon interactions.

K. A. Yasir and W.-M. Liu, *Rydberg Atoms in Cavity*, Springer Series in Solid-State Sciences 213, https://doi.org/10.1007/978-981-95-4571-1

Cavity Linewidth The spectral width of a cavity resonance, inversely related to its quality factor Q, which determines how sharply it selects frequency modes.

Cavity Mode Volume The effective volume over which the cavity field is confined. Smaller mode volumes lead to stronger atom-field coupling.

Cavity QED (Quantum Electrodynamics) The study of light-matter interactions in cavities where quantum properties of both fields and emitters (e.g., atoms or qubits) are important.

Cavity Pulling A frequency shift of the cavity resonance due to dispersive interactions with atoms or molecules, affecting lasing and stability in Rydberg masers.

Circular Rydberg States Rydberg states with maximal orbital angular momentum, offering long lifetimes and high sensitivity to external fields.

Collective Coupling The enhancement of atom-photon interaction due to many atoms coherently coupling to the same cavity mode, scaling as $\sqrt{N}$.

Cross-Kerr Nonlinearity A nonlinear optical effect where the presence of one photon affects the phase of another, enhanced in Rydberg-EIT systems.

Dark State Polaritons Quasiparticles formed by the coherent superposition of atomic and photonic excitations under EIT, with applications in slow light and quantum memories.

Dephasing Loss of quantum coherence due to interactions with the environment, thermal noise, or atomic motion, limiting the fidelity of quantum protocols.

Dipole Blockade A suppression of simultaneous excitation of nearby Rydberg atoms due to strong dipole-dipole interactions, fundamental for quantum logic gates.

Dipole-Dipole Interaction The long-range interaction between atoms due to their electric dipole moments, tunable in Rydberg systems.

Deterministic Photon-Photon Gate A quantum logic gate between photons mediated by Rydberg interactions or cavity nonlinearities, critical for photonic quantum computing.

Dissipative Phase Transition A quantum phase transition occurring in open systems, driven by the competition between coherent dynamics and dissipation.

Electromagnetically Induced Transparency (EIT) A quantum interference effect that renders an otherwise opaque medium transparent to a probe field, used in Rydberg nonlinear optics.

Fine-Structure Splitting The energy splitting between closely spaced atomic levels due to spin-orbit interaction, important in modeling Rydberg atoms.

Frster Resonance A resonant dipole-dipole interaction between Rydberg atoms when the energy levels of initial and final states match, enhancing interaction strength.

Fock State A quantum state with a well-defined number of photons or atoms, important in nonclassical light generation and Rydberg cavities.

Jaynes-Cummings Model A fundamental model describing the coherent interaction between a two-level atom and a single-mode quantized cavity field.

Kerr Nonlinearity A third-order nonlinear effect where the refractive index depends on intensity; in Rydberg systems, this occurs at the single-photon level.

Laser Cooling A technique to reduce the motional energy of atoms using laser light, often used to prepare ultracold Rydberg ensembles.

Lamb Shift A quantum correction to the energy levels of atoms due to vacuum fluctuations, relevant in precision measurements involving Rydberg states.

Long-Lived Coherence A property of Rydberg systems with extended coherence times, necessary for storing quantum information.

Master Equation A formalism for describing open quantum systems where the evolution of the system's density matrix includes coherent and dissipative processes.

Microwave Cavity QED The study of atomic and superconducting qubit interactions with quantized microwave fields in high-Q resonators.

Multistability The existence of multiple stable steady states under the same parameters, often observed in nonlinear Rydberg cavity systems.

Optical Bistability A nonlinear effect where the cavity transmission shows two possible values for a given input, enhanced by Rydberg interactions.

Optomechanical Coupling Interaction between light and mechanical motion, relevant when Rydberg atoms are integrated with vibrating membranes or movable cavity mirrors.

Photon Blockade A quantum optical effect where one photon in a cavity prevents others from entering due to strong nonlinearity, realizable with Rydberg atoms.

Photonic Crystal Cavity A cavity formed by a photonic bandgap structure offering ultra-small mode volumes and high Q-factors for integrating with Rydberg atoms.

Polaritons Hybrid quasiparticles formed from strong coupling between photons and matter excitations (e.g., excitons, atoms), central to cavity QED and Rydberg EIT.

Principal Quantum Number (n) Denotes the energy level of a Rydberg state; large n values lead to strong interactions and large polarizabilities.

Purcell Effect Enhancement or suppression of spontaneous emission of atoms inside a cavity, depending on mode structure and atom-cavity coupling.

Quantum Gate Fidelity A measure of how accurately a quantum gate performs the intended transformation, limited by decoherence and imperfect control.

Quantum Non-Demolition (QND) Measurement A measurement that extracts information without destroying the quantum state, possible in cavity-based Rydberg setups.

Quantum Optics The study of light and its interaction with matter at the quantum level, foundational to the understanding of Rydberg atoms in cavities.

Quantum Zeno Effect The inhibition of quantum evolution due to frequent measurements, explored in cavity QED experiments with Rydberg atoms.

Rabi Frequency The frequency of coherent oscillations between quantum states induced by an external driving field, such as a laser or microwave field.

Rydberg Dressing A technique where ground-state atoms are weakly admixed with Rydberg character, resulting in tunable soft-core interactions.

Rydberg EIT A version of EIT where the upper level is a Rydberg state, enabling strong optical nonlinearities and slow-light propagation.

Rydberg Polaritons Polaritons formed under EIT conditions with Rydberg atoms, allowing strong photon-photon interactions and nonlinear quantum optics.

Rydberg States Highly excited atomic states with large principal quantum numbers, exhibiting exaggerated electric dipole moments and long-range interactions.

Rydberg Superatom An ensemble of atoms within a blockade radius that behaves collectively as a single two-level system.

Single-Photon Source A device that emits one photon at a time on demand, often realized using Rydberg blockade and cavity-enhanced emission.

Spectral Diffusion The temporal fluctuations in resonance frequency due to environmental noise, affecting the coherence of Rydberg transitions.

Spin Chain A one-dimensional array of two-level systems interacting via dipole-dipole or van der Waals forces, often used to simulate many-body dynamics with Rydberg atoms.

State-Selective Ionization A detection technique where specific Rydberg levels are ionized for readout, enabling high-fidelity state discrimination.

Strong Coupling Regime The regime in which coherent atom-cavity interactions dominate over dissipation, necessary for observing vacuum Rabi splitting.

Superatom Model A theoretical simplification where a collection of atoms under blockade behaves as a single, collective two-level system.

Superradiance A collective emission process where the presence of other emitters enhances the spontaneous emission rate, relevant in Rydberg arrays.

Tavis-Cummings Model A generalization of the Jaynes-Cummings model to multiple atoms coupled to a single-mode cavity field.

Three-Level Lambda System A quantum configuration used in EIT, where two ground states are coupled to an excited state by two coherent fields.

Tweezer Arrays Optical traps used to manipulate individual atoms in programmable geometries, enabling Rydberg-based quantum registers in cavities.

Vacuum Rabi Splitting The normal-mode splitting observed in the transmission spectrum of a strongly coupled atom-cavity system.

Van der Waals Interaction A long-range interaction between Rydberg atoms scaling as $1/R^6$, dominant in off-resonant regimes and key to blockade physics.

Wigner Crystal of Rydberg Excitations A spatially ordered structure of Rydberg excitations arising due to strong interactions, observable in cavity-embedded atomic lattices.

MIX
Papier aus verantwortungsvollen Quellen
Paper from responsible sources
FSC® C105338

If you have any concerns about our products,
you can contact us on
ProductSafety@springernature.com

In case Publisher is established outside the EU,
the EU authorized representative is:
Springer Nature Customer Service Center GmbH
Europaplatz 3, 69115 Heidelberg, Germany

Printed by Libri Plureos GmbH
in Hamburg, Germany